Energies at the Farm

A Paranormal Investigation

George C. Bailey

The Cheeky Scribe

ISBN: 979-8-9947466-3-9
LCCN: 2026912404

Published in the USA by The Cheeky Scribe
Calumet, Michigan

Cover Photo: George C. Bailey
Author Photo: George C. Bailey
Cover Design: George C. Bailey and The Cheeky Scribe

First Edition

This Book is Dedicated to:

My wife, Gail English.
Without her help, this book wouldn't be possible.

My friends Caryn Schutte, Dan Braden, and my brother, Larry Bailey.
Their involvement was imperative to this story.

Carl Sagan, who guided me to keep an open mind during this process of investigation.

Contents

One

The Beginning

It was the summer of 1969 when I drove my dad's restored 1957 T-Bird to my friend Dan's home. His family's farm was just south of our high school (we were going to be senior in the fall) at Bullock Creek High School near Midland, Michigan. Little did I know that sunny visit would spark the beginning of the paranormal activity we experienced and discovered at The Farm referenced in this book's title.

Sister Ellen and me with Dad's '57 T'bird at the Bailey farm in Pleasant Valley, MI 1971

Without any expectations, I simply wanted to see the other side of this witty, urbane, humorous, yet grounded friend of mine. I thought a visit to his home would be appropriate. I had visited his home the previous summer when I rode my bike there, but this second visit was, well, more memorable.

The day of this second visit was perfect, weather-wise. The sun was high, unblocked by a cloudless sky. There was a light breeze and the temperature was in the mid-seventies. Dan and I again talked of things that interested us as high school boys, but none of those things stuck with me.

Except one.

"George," Dan confided, "Once on a farm where I used to live, I came out of the barn door to go to the house, and I levitated from the barn to the house."

I looked Dan straight in the eye and said, "Of course, you know I find that very hard to believe."

Dan also said that it was very much a day like the present day with a clear sky and full daylight.

Well then.

This was not exactly what I expected to hear from someone like Dan. Out of most of the people I had known up to that time, he was literally the most trustworthy and grounded person I knew. He had no detectable pretense or other socially accommodating behavior that would belie any unknown miswirement of his neural synapses. I filed his statement under the category of "curious and interesting," but doubtful for me.

Nothing else of consequence was said during our visit as I recall, and I left after a short time.

In the following summer and senior year in high school, Dan and I continued to discuss other issues that were important to us. No mention was made of his experience during that time. As life would have it, we graduated and parted ways, not keeping in contact.

About 35 years later, I contacted Dan out of curiosity to see what he was up to, since he and I were good friends in high school in the 1960s. We talked on the phone and then exchanged emails.

On August 6, 2005 we arranged a visit along with his wife, Jean, at their cottage near Alpena, Michigan. During one of our midnight chats under the clear northern Michigan sky, I asked Dan if he remembered telling me about his experience from our high school days. He replied that he had been waiting for me to ask him and, yes, he remembered it well.

He then added this summary of details.

For a time in the 1950s, his family lived on a farm in southwest Midland County, Michigan on Meridian Road. During this time, all members of his family frequently saw shapes and shadows of various human figures and glowing apparitions. It got so bad one night that Dan's parents had his siblings and him sleep in their parent's bedroom for the night. Dan then provided details on his levitation.

Dan stepped through a door at the barn and took a few steps towards the house with a full milk pail in each hand. He was then elevated about six to twelve inches and moved about two hundred feet towards the house in a gentle glide, then set down. After he was set down, Dan said to himself, "That was cool. Let's try that again." He went right back to the barn, but the aerial feat didn't repeat itself.

During our discussion, Dan told me exactly where The Farm was located. He said that his parents never spoke of the events in a way that could explain what was happening. Apparently they repressed talking about what was happening then because they didn't understand it.

My interest was piqued; I had a time, a place and specific events to piece together from Dan's early years.

Documenting the Unknown

This book documents several relevant aspects, from the time the Anishinaabe people lived in that area to the time the American settlers occupied this farm site in the 1850s. There are several aspects of this chronology with specific documented events and experiences with myself and the participants of these happenings which unveil many curious things regarding the Anishinaabe and European immigrant American settler activities.

In a nut shell, these are the documentations of the discovery of energies that have been experienced by myself, my friend Dan, our friend Caryn, my brother Larry and my wife Gail. These are chronicled events starting from the mid 1800s, when European settlers first established their presence on Anishinaabe grounds. Our documented investigation starts in 2004 when I began inquiring about my friend Dan's experiences from the 1950s. This book ends with my last visit to The Farm in 2023.

Throughout the course of those years between 2006 and 2023, I was able to enlist the help of some of my most trusted friends and family to explore and establish the facts and events of those energies that resided at this specific location. At the time, during these experiences, I was (and still am) a follower of Carl Sagan's teachings. He was an influential teacher who stressed the scientific method of collecting and evaluating facts with an open mind when investigating the world around us. "Extraordinary claims rely on extraordinary evidence" is one of the many points of his teachings.

Most of the events chronicled in this book are the interaction of energy sensitivities and researched facts. The timing of these events and facts are significant to establish the importance of what happened during our experiences. All of us entered into this phase of fact-finding without preconceived notions of what we might find. I was the chief driving force to coordinate our meetings at The Farm and to extend our outreach to others who had previous experiences relating to The Farm. I did this to find the facts of what others had experienced, so that a somewhat comprehensive compilation of information could be done. I am presenting these events in a way that shows we relied on the sensitivities of gifted people in a manner that was subsequently confirmed by researched facts.

As we unraveled the story of the tragic events that befell one of the early families who resided at this farm, it became evident early on that there were other energies that resided there. It appears that, in real time, the second generation of early settler families did not know of the Anishinaabe connection to the land. There are apparently no written documents of any direct knowledge of the Native American presence by the settler descendants.

There are five main people in this story who worked as a team to share their strengths to find out what was going on in the near recent past and what had happened further in the past.

Being the author of this work, I volunteer myself for first scrutiny.

The Energy Detectives

I, **George C. Bailey**, recognized that this was an opportunity to learn and discover things about energy sensitivities that I had little knowledge or experience with. I learned from the teachings of Carl Sagan earlier in the 1970s to form my skeptical outlook on many things.

I liked the need for evidence to establish world order in my mind. Although I eventually became aware of some sensitivities to energies, I originally blocked out my need to be more cognizant of them. I was skeptical when Dan mentioned his experience at The Farm in 1969. I was also quite naive about life in general then because I was 17 years old, and I considered myself a slow learner. I didn't know the value of asking questions about things I didn't understand.

By the time I reconnected with Dan in 2004, I had more experience in questioning the unknown.

Danny Braden is key to this story, because he is the main reason I started researching this story. He, as it turns out, is quite sensitive to energies. He appears to me to be sensitive to seeing energies, not necessarily feeling them. Dan, as I mentioned earlier, is one of the most level-headed people I know, then and now. He has always been a solid friend, and I trust him implicitly.

He is a person who has an excellent grasp of the material world, and I have asked his advice on many things related to machines and how they work. He is a great person to know.

Danny Braden 1970 George C. Bailey

Caryn Schutte is very sensitive to energies and has been all of her life. She, as she has revealed to me, is more of a feeling type of sensor, compared to Dan. She has also stated that the energies themselves have always felt free to come to her when they need something closed out on their way to their final resting place.

I met Caryn in 2005. She and I started visits to The Farm that same year, and she has been a key part in our investigation.

Larry Bailey is my brother, and he is also quite sensitive to energies. His type of sensitivity is more of a visual type of gift. He has related to me that, prior to this investigation, he has had a strong sensitivity to Native American energies, although he has had many sensitive contacts with energies throughout his life.

I am married to **Gail English,** and she has had sensitivities throughout most of her life. We had a memorable experience during one of our visits to The Farm in Chapter 3. Although her experience at The Farm was a tactile one, she also feels many energies much like Caryn does.

From my descriptions of the people involved with this investigation, one can see that there is a difference in how these people are aware of the energies they have encountered. Feeling versus visual sensitivities indicates to me a different level of awareness regarding abilities.

Dan and Larry appear to be more of the 'visual' strength, while Caryn and Gail appear to be of the 'feeling' strength. Is it a matter of gender that separates their strengths between Dan and Larry compared to Caryn and Gail? Either way, Caryn had a visual event reported in Chapter 5.

In later years, I have sensed energies both from a visual and feeling awareness. I suspect that people develop sensitivities between the feeling and visual forms of awareness, similar to energies when they crossover from their living bodies to the "other side." I have been aware that people's energies may 'heal' when they die and crossover to the "other side" to attain a higher vibration. It appears to me that it is similar in context to what I experienced when I not only sensed energies but saw them also.

I have seen a female figure swathed in draping white cloth cross on our property (also seen by Larry and Dan at separate times in daylight). I have also seen numerous energy forms in the shape of a punctuational comma about four inches tall in full sun light on our property (when I returned from downstate). Larry and Gail saw them as well.

This event was a reality check that energies were a common reality. Dan and I have seen pinpoints of intense light on our property during nighttime. He and I saw them repeatedly on our shore of Lake Superior. However, I haven't seen them since I first saw them with Dan.

Gail and I have had many ocassions where we have felt energies on our property. She and I spent a night in St. Ignace, Michigan in 2013, and our night was intense with energy activities. The historical house next door was purported to have a history of energies. One thing that seems to have a common link to most of these events is, when a new person arrives at a location, there is an increase of energy activities for the person arriving.

One needs to have a critical eye regarding the energy events I have experienced. Carl Sagan taught the scientific method of qualifying that "extraordinary events need extra-ordinary evidence" through reproducing experiments. These events we experienced don't fare well when I or others try to prove or disapprove these energy events by reproducing them. To recreate these events under a scientific test would be difficult to do for this layman. Those tests are probably not available to definitively prove or disprove them. I like to use Occam's Razor for one of those tests. The Occam's Razor test is simple in that it states that a hypothesis about a fact or event probably has the simplest reason why it has occurred. It is similar to the cliché that if it looks like a duck and sounds like a duck, it's a duck.

These observations about my energy experiences appear to be a duck, if you will, unless some other evidence would change my view. True, these stories about my energy experiences are anecdotal (see Chapter 5 for the exception), but they are nonetheless my history of observations. I need to allow the supposition that I am either delusional about this investigation, or they could be explained by quantum physics or a third effect.

The scientific method may eventually confirm that quantum physics is in play, or they're at least there partly due to the reason of why these energy events could happen at all.

Two

The First Visit to The Farm

On September 20, 2005, my friend Caryn and I made our first visit to The Farm. I invited her because she had indicated to me that she was sensitive to spirit energies and was interested in exploring The Farm with me. Since I felt I had almost no ability to communicate as she could, we arranged our first visit. I didn't know what to expect from this visit, but my intuition regarding Dan's levitation experience decades earlier had elements of unexplained reasons that *must* have a reason. We were essentially going into this event with no preconceived ideas of what we would discover. Caryn mentioned earlier about her gift of sensing energies, but I had no previous experience of her ability to do so.

The day was a picture perfect early autumn day in the mid 70s without a cloud in the sky and no wind, similar to when Dan first told me about his levitation in 1969. I had previously driven by The Farm a few weeks prior to make sure I could find it after Dan described the location.

The following is Caryn's description of what she experienced when we explored The Farm site.

> *George and I were deep in conversation when he all of a sudden pulled into the driveway and said, "Here is The Farm." I quickly explained to him that I never call a spirit to talk to me but would open myself up for a communication mode. George already knew this about me. I then went on to tell him that he was to leave me alone, and he do his thing* (taking pictures), *and I would talk to him when I needed to. If I had a dialogue*

going, it may be internally, or I may voice it for his benefit, but I would feel the situation and go with it as it felt natural.

I then opened the door of the vehicle and immediately felt a presence, never before to that magnitude. My chest was full and warm, and I know there was going to be a contact. I went over to the foundation of the house and immediately knew I was not alone.

Sometimes I actually 'see' the entity, and sometimes I 'sense' it. This was a sense contact. There were two children running around the outhouse, the boy about 9 and the girl about 7. It was very clear the mischievous brother was going to lock the sister in if he caught her. I did not know it was an outhouse and later asked George to check it out, as I was not able to walk well that day. I found myself laughing out loud as the children ran around and then 'Dad' communicated with me stating they are always causing trouble. I replied with "at least they are normal children." I knew Mom was there and was very skeptical of me. When I reassured her I meant no harm, she let me know they were happy there and would continue to stay there.

This came about because when a spirit comes to me, most of the time they are asking my permission to go to their better place. This family was clearly happy where they were.

I sensed another entity, thinking it was a baby, but did not feel clear about this. At this time it was made clear to me this family died on this property; the house itself was not important, but the land was. They were of pioneer time.

The family understood why I was there, and, although they did not say "Dan's family" or anything of that nature, I knew we were talking about the fact that Dan levitated. It was communicated to me that that family was well liked because they seemed to accept their presence and not try to get rid of the spirit family. The levitation was a joke, and they enjoyed playing jokes on everyone, especially Dan's family because they were so accepting. They never meant any harm.

At this time George came to me, and I gave him a brief rundown of what had transpired.

He then informed me his camera was acting up by the barn. I asked him to help me walk there, so I could see if there was some other entity. I looked around the barn as George was explaining about the milk parlor and where Dan probably was, and I knew the children were in the loft, so I knew they were the only ones present.

As I stood outside the barn and waited for George to get the vehicle, I closed my eyes to make sure I had the whole sense and did something I never do: I asked for clarity of who else was there. The baby I previously wasn't sure about was indeed the presence of Grandma and a baby.

I have never had an experience so intense and strong, and I am not sure why it happened that day, but it was good, positive and comforting.

Analyzing the Data

After I read Caryn's description of our first visit, my thoughts about Dan's levitation event were one of the first I questioned. I have a hard time believing why Dan defied conventional physical laws when he was lifted with two full milk pails and transported 200 feet before being set down gently. This seemingly impossible action still does not make sense to me. The physics of that purported action defies common sense for most people, including me.

Dan, as I described earlier, has no apparent need to purposely deceive me. I am convinced that, given my past experience of knowing him for the four years before we graduated high school and the 20+ years I have known him since we reconnected in 2003, he is convinced it happened to him. Given his age at the time (about six years old), and that he had two filled milk pails, the total weight was probably at least 100 pounds. That weight being levitated a foot from the ground and transported 200 feet, shows the physics of levitation involves a significant amount of energy.

Another possible explanation of his event could be a hallucigenic phenomenom. I have a good amount of respect, and I am leery that the human brain can create reality where no apparent reality exists in our real world. Did Dan have a convincing hallucination when Caryn related what the energies said to be the reason for his levitation? Were both Dan and Caryn being deceived with an alternate perception of reality from their experiences?

I doubt it.

It appears unlikely that they have both been deceived about the levitation. Still, what energy force explanation could have allowed such an event to happen? Are there quantum physics involved that allow crossover actions to enable energies from past human entities to think it would be a good joke to play on people on our side of reality?

After this entire investigation, I suspect that quantum physic reality interacts to a certain degree with our perception and laws of reality. The levitation at the specific time that Dan experienced it, is not practical to reasonably repeat and test. Could Caryn's contact with the energies that supposedly carried out the levitation be repeated?

Possibly so, but as we will learn later, the family of energies have apparently moved on to another level, so I think her method couldn't be reproduced. I am surmising that further discoveries about the quantum physics crossover affecting the physical dimension of which we are most knowledgeable will be forthcoming: probably after I, and the participants in this investigation, pass on. I suspect that a parallel universe is quite close to our universe that allows interactions to crossover due to the nature of those quantum physical actions.

No wonder these currently unexplainable events happen while we don't understand or have a reason.

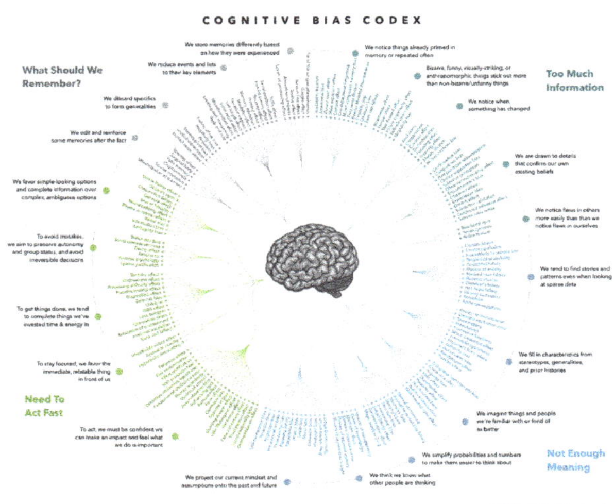

The possible way my brain may have misled me

Identifying the Entities

There is a point of circumstantial time-related evidence that I want to point out regarding our first visit. Caryn had sensed that a mother and two children made themselves apparent. She also sensed a father, a baby and a grandmother. In Chapter 5, the mother and children are identified: Margaret as the mother, Donald as the male child and Frances as the female child. Apparently, all three of them died tragically in a house fire that occurred in 1928. Frances was 16 when she died, and Donald was 3.

There is a discrepancy in the ages of the children as described by Caryn in her first visit.

In my view of the events, Caryn identified the entities as a mother and two children. In addition, she felt a father, identified later as Thomas Dean, and a grandmother, possibly Josephine Doutre along with a baby, who isn't identified in later research. Perhaps the baby died in infancy, but there is no evidence in the records that the infant existed.

A grandfather (Frank Doutre), grandmother (Josephine Doutre) and two other older children are accounted for in public records. The grandparent's other children lived into adulthood and, apparently, their energies were not present at The Farm during Caryn and Larry's visits. Although Caryn wasn't able to establish the names of any of the family names during her first visit with them, she does establish that there is a distinct number of family individuals. This, to me, indicates circumstantial evidence for this view of my understanding, but after the events that occurred later in this story, I think there is a strong case for energy interactions with Dan, Caryn, Larry and Gail.

While none of these facts from our first visit stand as strong evidential proof, I think the evidence that the unusual events (to us) have occurred in the past at The Farm are warranted.

Seven days later, I sent Dan an email telling him what we experienced.

Hi Dan,

A few days ago, my friend Caryn and I visited your old farm house where you used to live in your earlier days as you described to me. I asked Caryn to go with me because we both have an interest in the experiences you shared with me in our high school days and from when you invited me to Alpena to your cabin a few weeks ago.

I took nine pictures, which I have attached, that you probably recognize. The first picture is of the southeast foundation of the house looking towards the east to the barn. You will also recognize the picture inside the barn looking back west to the house foundation, a picture inside the attached south shed/milking room and from the shed/milking room towards the house foundation. As a curious note, when I took the picture through the broken door towards the house foundation, I turned my flash on, but before and after I clicked the camera, I got some unexplained distortion in my view finder that I have not had before, nor since.

Caryn is gifted with an ability that she was kind enough to share with us that I think you will find interesting. Following is her account of what she experienced when we visited your old farm.

I included Caryn's previous summary in the email to Dan. The rest of my email follows:

Well Dan, that was it. Do you have any insights or comments you would like to add? I plan on doing some research on the pioneers that settled the farm. The farm is bounded to the north by Bullock Creek, how coincidental, and is located in Section 28 of Mt. Haley Township in Midland County. Pretty interesting, huh? Feel free to share this with your family if you feel it is appropriate, for I bet they have some insights and experiences that I would like to hear.

Later, George

Eight days later, Dan replied.

Hi George, I have been reading your e-mail over and over and over for the last week. It has left me speechless. I wish I had been there with Caryn to ask some questions of her and the presences. She sounds like a remarkable person. Any further info you come up with about the farm would be of great interest to me. I have been wondering for years just what was there. I was able to sense something but never had the gift to know what it was. If I had this information years ago, it would have made a huge difference in my life. Experiences from that part of my life formed how I think and feel about things all through life. I never recognized the happy people of a past life around me. All we ever saw were dark shadows and glowing figures. It could have been so nice to interact with our co-inhabitants. I wonder if there are others there?

Perhaps a darker presence. I am not sure everyone there was happy with us or our presence. I felt that at times they were trying to scare us away.

I want to thank you and Caryn for taking the time to go to the old farm to do the research. I am stunned at what you both experienced. I had no doubt that you would experience something! I was just not ready to hear for real what I already suspected. I am glad it was a positive experience for Caryn. I was afraid it could have been a very dangerous situation for her. It takes a lot of nerve to go to a strange site and experience who knows what. My hat goes off to her, she is one brave lady. I would hope to meet her some time.

Take care my friend, Dan

Later that day, I replied to Dan.

Hi Dan,

Yes, it was a rather remarkable experience, even for this Carl Saganist. Caryn would love to meet you when the schedule permits, and we will try to plan on it. As much as I have been using my strict scientific outlook on the world around me, it is harder and harder to ignore the fact that three of the more influential and most trusted people in my life: you, Caryn and Larry, have had a lifetime of experiences of this calibre. Unfortunately for me, it appears, until I have a similar experience, I will continue to doubt the existence of these events due to my self-indoctrination of needing facts, reproducible results and measurable data. Of course, I just may not be sensitive to these phenomena and never will be.

As an offhand thought, I would like to possibly follow up with another meeting at The Farm with you and/or Caryn. Of course, I will discuss with Caryn this scenario, since, as she stated earlier, she does not seek out these experiences, and I don't intend to pressure either of you to pursue this any further. If either of you see no good coming out of any more visits or other actions by me or us, I certainly will not pursue it.

I hope I have not dredged up any bad memories or experiences that would be

> *best left to fade away. If I am stirring a pot that is best left alone, I will not disturb this totally fascinating situation and condition that just begs me to investigate further.*
> *As always, I am very interested in your thoughts about this subject and any other slants you may have on it.*
> *Later for sure, George*

After these email exchanges, Dan had reservations about our active inquiries at The Farm. In an email he said:

> *I must say I worry about every time someone goes to the place on M-30* (now known as Meridian Road at the time of this writing). *I still think there is more there than is being shown. I think not a good thing. I am just glad that everyone has come away unharmed. The more I think of it, the less I think I will ever set foot there again. I may be wrong, but I think not. I hope I am. It is such a powerful place.*
> *Please be very careful.*
> *Best wishes, Dan*

That statement by Dan did not deter my need to continue the investigation (despite my statement to not 'stir things up') because I didn't feel insecure with the possibility of 'bringing home' any energies that might be detrimental to my well being. Actually, I was pissed off by the thought that a belligerent energy would try to invade my space to harass me.

As I reflect on this reaction about a presence possibly invading my space, I think my aggressive attitude regarding a possible threat may be related to an insight that Caryn had when Dan was visited by a screaming apparition (Chapter 5) a few years later. At that time, Caryn advised Dan to tell all energies that they were not welcome in his life, and they should not bother him. After Dan did that, he did not experience very many visits. I am wondering if my attitude may have accomplished the same deterrence.

The Farm when we first visited

In an email Caryn sent in 2023, she reflected how her first visit to The Farm affected her personal outlook from sensing the energy entities.

> *One of the profound moments that stood out for me from The Farm experience was being validated in my ability to communicate with the other side. This was the first time ever. I have had many visits from the spirits of many people I have known as they die.*
>
> *Typically, they are wanting me to go to their family members with a message, or maybe just to be there in their time of grief or sometimes to say good-bye to me. I love old cemeteries and, when visiting, have always known the presence of an entity. Loved ones have come back in many forms to visit me, give me a message or support me in a time of need over the years. I have helped chase an evil entity out of many situationsover the years. This has met me with sharp criticism from many because, according to the Bible, I am dealing only with evil, and that makes me evil. Many times I have wondered if I am just crazy even though, deep down, I know what I know.*
>
> *Back to The Farm, upon my first visit I came to realize that there had been a family that had a bad house fire and that some members perished. They made their presence known to me. I described them to George and Larry in detail including the accident, the ages and who was left. The next time I saw*

Larry, he brought me a newspaper article that described everything I had said with great accuracy. Finally, I had validation.
Another incident that had a profound affect on me was the one day we were there (The visit on June 25 and 27, 2006 Chapter 5) *and the "mother" was sitting next to me on the foundation as I was talking to George and Larry about what was going on that day. I felt her shift her position as she looked at Larry, and I was very aware that someone was stroking his cheek. It was her. She told me he was so handsome but did not understand why he was clean shaven, not common in her time. At that moment, I realized how the entities long for the feeling of being alive.*
I have not been scared since, when I am somewhere, and they try to 'enter' my being. I understand, so I can use my compassion now while standing my ground, so they are not allowed to take over my body. George witnessed this once when we were at a cemetery (in Ithaca, Michigan, my first hometown) *completing a task on his ancestral quest. He said the whole stature of my body changed and my face became almost contorted as if it was another's face.*
That was the most extreme incident I had like this, and it was before I had the realization mentioned above. Instinctively, I knew to ground myself and be stern.

That visit to the Ithaca, Michigan cemetery in 2006 was the time Caryn said to me that there were many troubled energies there. I had several experiences in our house at 121 Arcada Street in Ithaca, Michigan during my youth before 1960. I frequently had dreams that fearful energies were in the drain of the bathtub. It was a scary energy, but I had no visions to relate to that terrifying fear.

This first visit to The Farm opened many questions in my mind, such as: Why was there activity that apparently had not been explained by people not involved with the family? Of course, the events that happened to Dan's family and other people associated with The Farm energies happenstances were not common topics to discuss amongst people at that time. These happenings, as other events in day-to-day life, didn't make sense to the residents that were in knowledge of those happenings. As with many things in life then, they probably considered them unknowable or matters that people considered as works of Satan or other untouchable subjects.

In an interview with Hazel (one of the children who experienced the fire in 1928), she said that their family didn't experience any energy happenings. It is apparent that, at least with Hazel, she didn't have the sensitivity to the energies. Likely, there were not any energy events that were made evident until Dan's family lived there. Possibly, after this book comes out, some residents or their descendants may have more history to provide for this investigation. The Native American energy descendants that are part of the portal location (Chapter 7) may have some insights to their ancestor history.

Three

The Second Visit to The Farm

Gail and I visited The Farm the last week of February, 2006 because she was interested in our investigation. Although she had passed The Farm countless times before, as I had done, she didn't remember anything memorable about this typical abandoned farm property as it was a common sight in our Midwest scenery.

We did a quick tour of the premises and were about ready to go on our way, when we walked to an outbuilding as seen in the accompanying picture. We were holding hands, with me on the right and Gail on the left, when we peered into the doorway because Gail said she didn't want to float away before we approached the outbuilding, apparently referring to Dan's previous levitation experience. As we stood just outside the doorway, and leaned in for a look, Gail jumped and said, "Someone just put their arm around my waist!"

It certainly wasn't me because I was holding her hand. Needless to say, this was a startling event that we weren't expecting. We didn't go into the building like I expected to do. We left The Farm shortly after that. I didn't take any pictures of the area due our uneasiness of this unsettling tactile experience.

Throughout this book, I look at the events that occurred with a view that most of them may tell an interesting story, but what evidence exists that supports the findings? The evidence for this event relies on an anecdotal retelling of what happened to Gail as told by me. The practicalities of repeating this event are not realistic because too much time has elapsed, and the location of the building and the energy that caused the purported event are not present anymore, as you will see later (Chapter 12).

Gail sent me the following message on February 28, 2006 recalling her feelings a few days after this second visit.

> *I understand Dan's concern and trepidation to have the experiences he did as a young child; it must have been so frightening. I would imagine his parents' and siblings' reactions were anything but calm or in any way enlightening as to the explanation of the appearance of entities. I felt no fear but I, at this point in my life with more understanding and a different perception of fear itself, really can't think of anything I fear or feel uneasy about. The jolt I felt at the entrance of that outbuilding seemed to be a presence making itself known. A bit startling, but not negative. I feel no need to try and talk Dan into going there if he doesn't want to, just wish someone with more knowledge could share it with him.*

Gail's descriptions of her literal feeling of the arm around her waist showed me that the energies do not appear to be malevolent, as Caryn confirmed from our first visit the previous year. It appears that they are 'playful' as described by Caryn, and they appear (in hindsight) to be from the family of Margaret. As we will see later, they appear to not be Native American energies as shown in Chapter 7.

The outbuilding where Gail felt an arm around her waist

Four

A Spirit Session

In March of 2006, Gail and I participated in a meeting with a medium who was able to communicate with energies who spoke through her. Due to circumstances of a sensitive nature, I am not able to mention the medium's name. However, this medium was able to facilitate these energies through spoken words as the medium turned into the energy's mouthpiece. I documented the recording of this session with the intent to find evidence of manipulation of the facts found during the session for posterity. There were some instances where the medium could not have known details that were only known to me. Although there were some details that possibly could have been faked or found out by the medium beforehand or through her intellect, I am convinced by the evidence I collected, that the medium has a gift of sensitivity to communicate with energies.

The following paragraph is an excerpt from the composition I did in 2006 about the "A Spirit Session" event. Although there are several fascinating details from this session, this excerpt provides some insights in the understanding about what we found at The Farm.

> *When Hiram left,* (a spirit energy prior to the Native American visitor in this session) *an un-named Native American came to talk about the farm where Dan Braden used to live on Meridian Road in Midland County, Michigan when he was a child. The native spirit said that Dan's farm home was the site where the natives burned the bodies of the deceased in funeral pyres. The native spirit said that the spot where the spirits lifted from the pyre was a chosen site by the Native American spirit guide/shaman/medicine man who lived with the resident Native Americans in earlier times,*

> *because there is an upwelling type of conduit energy flowing upwards from the ground up to the spirit world.*

During this session, the information about those particulars of Native American funerary practices was not known to us investigaters. Also, we didn't know that the location under the northeast corner of the barn was the place where the past energies of the native residents apparently ascended. Larry gives specific dimensions of the portal at The Farm in Chapter 7 when he describes it at The Farm on August 10, 2008.

Gail sent an email to me describing what she shared with Dan during the spirit session, referring to The Farm. Gail continued her email to Dan.

> *I'll review the information regarding The Farm we got from the session. The land was used for spiritual purposes by Native Americans prior to the time that the Doutre* (and Dean) *family lived there. The Native Americans believed that* (The Farm) *land held a sacred area conducive for the spirits of their dead to rise and become one with the Great Spirit. They burned the bodies of their dead in a ceremony, so this could take place. That land was and still is a very* (positively) *powerful place for spiritual communication, hence the 'truth quakes,' as the medium called them, felt by energy sensitive humans who visit there. Nothing negative, but I can certainly understand your apprehension, Dan after experiencing what you did as a young child. I would have reacted the same way. I can't imagine how my parents would have handled those occurrences with us. At this stage of my life I'd be curious and delighted, I think, because I have had more paranormal experiences and education. George's great great grandfather, Hiram Stanton, also visited us during the session much to George's delight and gave him information to further his genealogy quests.*

Below is a photo taken at the Assinins Community Cemetery in Baraga County in the Upper Peninsula of Michigan. I visited the cemetery in 2023 in order to get a better picture of the sign, but apparently, it has been removed. This cemetery is part of the Anishinaabe (Indigenous People) that may describe the customs of the former residents of the site of The Farm. On this marker, there is a description of how food and other items were left at the burial site.

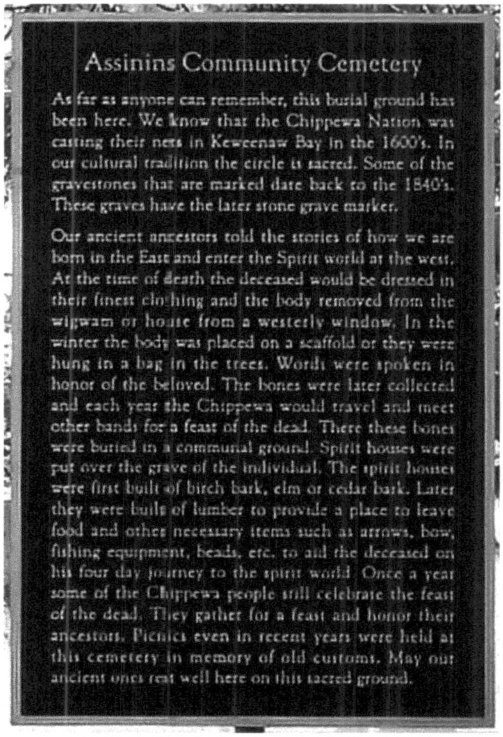

In Chapter 11, I took pictures of an offering at the portal location inside the barn. I don't know who placed them there, but it appears to me, that this was an offering of deer fur and skin for the Native American energies who are familiar with the energies residing there.

Five

The Third and Fourth Visits to The Farm

Caryn, Larry and I visited The Farm on June 25, 2006 and returned two days later for a follow-up visit at night time about 9 p.m. This was Larry's first visit, Caryn's second and my third. I invited Dan to join us, but he was reluctant to visit because he felt a lack of control regarding bringing any of the energies 'home.' Of course, I respected his decision to protect himself and possibly others. I did think that he was missing an opportunity to find out some cool things about his past, but I wasn't going to pressure him with the same need as I had to find explanations.

As I mentioned earlier, I did not feel the same need to protect myself. I still had the confidence that an unknown possible threat wasn't going to prevent me from finding the reason behind these prior events. I realized later that my stubbornness to plow ahead may have ended badly, but in retrospect, I did find the reason why these events happened. The following is Dan's message declining to visit with us.

> *Hi George,*
> *I will decline your kind offer to join you for a tour of the old place. After much thought, I have decided to let sleeping dogs lie. As I had mentioned, I feel the spirits are not locked into or restricted to just that space. The last thing I want is to bring one home with me as a lifelong guest. I was lucky enough to escape them once or twice. I do not want to press my luck. I have had several experiences of several types and in several places. In each experience, I have never felt in control. I don't like that much. Just outside my comfort*

zone, way outside. I will take on life's challenges as I get to them, but I am not willing to go look for more unless there is a need. This is not a case where I feel I want to go right now. Sorry it took so long to reply, but I really needed to think this over. Good luck with your visit, and keep me posted as to what you find. Talk to you soon.
Take care my friend, Dan

Our third and fourth visit to The Farm is best described in Caryn's words, as she recounted events from Sunday afternoon starting about 1 p.m. and the following Tuesday evening starting about 9 p.m.

My second visit to The Farm was made with an air of unknowns for me. There was a different atmosphere presenting itself. I was not sure what Larry's presence would bring to the whole experience. He went not because he communicates, per se, but because he wanted to see if he could come up with any visions of what went on at The Farm in former times. I was anxious to see if I could draw off his energies.
When we arrived, I stepped out of the vehicle and immediately felt the presence of the entities. I walked over to the corner house foundation, which seems to be the place of communication, and as I sat to still my spirit within myself, I felt an unusual confusion amongst everyone (the entities). *At first I was wondering if it was because Larry was there, or, perhaps it was George milling around, or maybe they sensed the uneasiness I felt.*
I took the time to assure all that Larry was fine, George was here before and yes, they were agitated, but not because of them. I was then presented that there was confusion between the 'back' and 'front.' Let me define this right now. I consider the 'front' to be the family that I have communicated with before and the 'back' is a mystery to me as to who it is (this would be revealed in Chapter 7). *That was never revealed to me. I know when I crossed the 'boundary,' as I will explain in a moment, I felt an unrest, an agitation or a closer description, a despair* (this corresponds with the portal discovered in Chapter 7).
The birds helped me to find the boundary, the line that seemed to separate the front and the back of the entities that are present on The Farm. (Earlier

at lunch, Larry was talking about the progression the Native Americans feel the spirits take as they make their way to their final resting place. He remarked somewhere in the conversation that birds were of one of the higher places (of energy levels), and that is why eagles were held in such high regard).

During the confusion period I felt when I first arrived at The Farm, I noticed that there were two starlings sitting in the willow tree off to the left of me. They were perched in branches facing each other about three feet apart. I thought of Larry's words and stilled myself to listen. The front bird seemed to be telling me, "don't go back there, it is not good!" And the back bird seemed to be edging me back, telling me in the process that things had to be settled.

This unrest was not going to work. I took one step at a time, working myself back very slowly. With each step the front bird would initially get very upset but calmed down once I seemed to stay put for 30 seconds or so. I continued until I discovered that the boundary was the trunk of the willow tree. I walked back (beyond the boundary) casually for a few more steps and, at that time, felt as if an entity was perched on my shoulder and wanted a ride back up front. I went up (back to the front) and was aware of a discussion going on between front and back entities. It was almost as if it was spoken in foreign tongue. A sense of calmness seemed to take place, and again, I felt as if I had to carry the back entity to behind the border, so that is what I did. (After I did) I stood for a very long time (a few minutes) to make sure I was grounded, and there was no one still with me as I felt the feeling of despair very strongly. I took a few deep breaths in and blew out slowly until I felt at peace and alone.

Sometime after this, Grandma asked if it was all right for their family to leave. I replied with "They must feel right doing what they wanted." After the return visit on Tuesday, I am fully aware they will not be going anywhere. They are content. (This changes in Chapter 7 when Margaret crosses over, and the family is not at The Farm. Apparently they crossed over prior to Margret's crossing.)

Back in my apartment in the middle of the night after I used the bathroom, I felt there was someone with me. After I laid back down I felt scared. I have never been scared before. I tried to go back to sleep, and after getting irritated with the whole way I was feeling, I commanded "Either reveal yourself or leave." I was then at peace and alone.

Larry, George and I felt that a return trip during the still hours of the evening might prove beneficial, so we agreed to meet about 9 p.m. on Tuesday and stay for about four hours. I was called to the 'back' as soon as we arrived. I walked back and seemed to be thanked, but there was a crow that was screaming at me as if to say, "you really screwed things up!" I have never seen a bird take off as sharply around a corner (not a physical corner, but a sharp turn) as this bird did. I feel the 'back' for the most part, felt I served as a calming influence or peacemaker, as needed, with the exception of this one entity. After that, I would not cross into the back (this visit). I did not feel welcome nor secure to go there.

George, Caryn and Larry June 27th at The Farm

Caryn continued.

When Larry arrived (Tuesday evening) he said "Caryn, you are good!" and he handed me a copy of a newspaper article with "my family's" story on their tragedy and deaths in that house. To see faces of whom I have been communicating with has been so overwhelming. Two months later, not a day passes without me getting the pictures out to look at them. In the newspaper article, there is the Mom named Margaret, a daughter and a small son who died in a fire as I described the first time I visited The Farm. This is profound for me, and I feel validated in my communications for the very first time in my life.

Burns Fatal To Three

Victims of Mt. Haley kerosene blast are pictured above. This cut was made specially for The Republican from rather poor snap-shot photos, obtained from the family. At the left is 3-year-old Donald, right above Mrs. Dean and below Frances. See story in another column.

MT HALEY MOTHER, TWO CHILDREN DIE AFTER EXPLOSION

Kerosene Used To Start Fire Wrecks Thomas J. Dean Home Near Jam Early Tuesday

BOY DRAGS SISTER OUT; PUTS OUT FIRE

[illegible]

Lad Tries To Save Mother

[illegible]

Mother Braves Flames To Save Infant

[illegible]

The newspaper story – first two columns

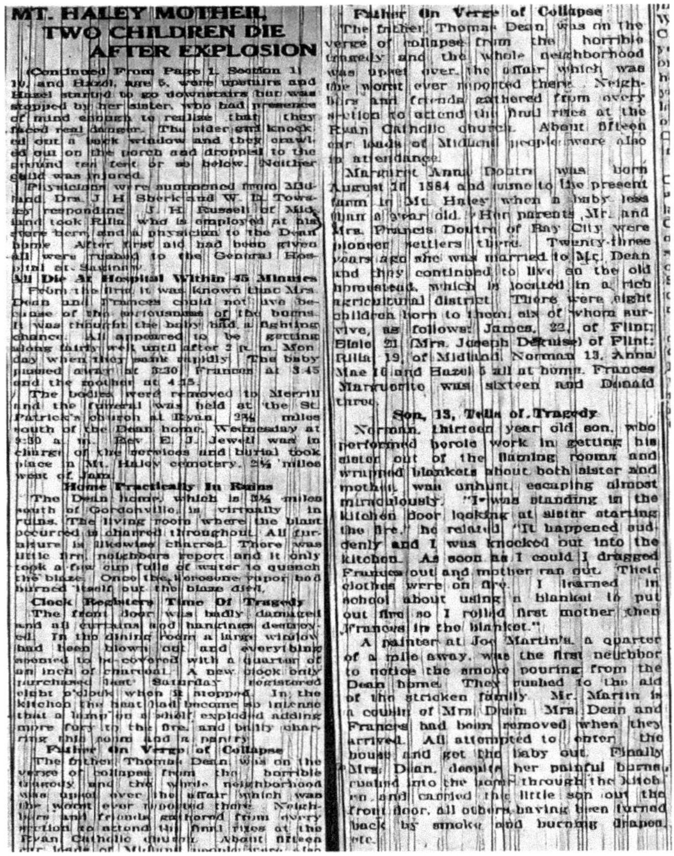

MT. HALEY MOTHER, TWO CHILDREN DIE AFTER EXPLOSION

(Continued From Page 1, Section 1)

10, and Hazel, age 5, were upstairs and Hazel started to go downstairs but was stopped by her sister, who had presence of mind enough to realize that they faced real danger. The older girl knocked out a back window and they crawled out on the porch and dropped to the ground ten feet or so below. Neither child was injured.

Physicians were summoned from Midland, Drs. J. H. Sherk and W. E. Towsley responding. J. H. Russell of Midland took Rilla, who is employed at his store here, and a physician to the Dean home. After first aid had been given all were rushed to the General Hospital at Saginaw.

All Die At Hospital Within 45 Minutes

From the first it was known that Mrs. Dean and Frances could not live because of the seriousness of the burns. It was thought the baby had a fighting chance. All appeared to be getting along fairly well until after 2 p. m. Monday when they sank rapidly. The baby passed away at 3:30, Frances at 3:45 and the mother at 4:15.

The bodies were removed to Merrill and the funeral was held at the St. Patrick's church at Ryan, 2⅜ miles south of the Dean home, Wednesday at 9:30 a. m. Rev. E. J. Jewell was in charge of the services and burial took place in Mt. Haley cemetery, 2½ miles west of Jam.

Home Practically In Ruins

The Dean home, which is 3½ miles south of Gordonville, is virtually in ruins. The living room where the blast occurred is charred throughout. All furniture is likewise charred. There was little fire, neighbors report and it only took a few cup fulls of water to quench the blaze. Once the kerosene vapor had burned itself out the blaze died.

Clock Registers Time Of Tragedy

The front door was badly damaged and all curtains and hangings destroyed. In the dining room a large window had been blown out and everything seemed to be covered with a quarter of an inch of charcoal. A new clock only purchased last Saturday registered eight o'clock when it stopped. In the kitchen the heat had become so intense that a lamp on a shelf exploded adding more fury to the fire and badly charring this room and a pantry.

Father On Verge of Collapse

The father, Thomas Dean, was on the verge of collapse from the horrible tragedy and the whole neighborhood was upset over the affair which was the worst ever reported there. Neighbors and friends gathered from every section to attend the final rites at the Ryan Catholic church. About fifteen

Father On Verge of Collapse

The father, Thomas Dean, was on the verge of collapse from the horrible tragedy and the whole neighborhood was upset over the affair which was the worst ever reported there. Neighbors and friends gathered from every section to attend the final rites at the Ryan Catholic church. About fifteen car loads of Midland people were also in attendance.

Margaret Anna Doutre was born August 18, 1884 and came to the present farm in Mt. Haley when a baby less than a year old. Her parents, Mr. and Mrs. Francis Doutre of Bay City were pioneer settlers there. Twenty-three years ago she was married to Mr. Dean and they continued to live on the old homestead, which is located in a rich agricultural district. There were eight children born to them, six of whom survive, as follows: James, 22, of Flint; Elsie 21 (Mrs. Joseph DeKruise) of Flint; Rilla 19, of Midland; Norman 13, Anna Mae 10 and Hazel 5 all at home. Frances Marguerite was sixteen and Donald three.

Son, 13, Tells of Tragedy

Norman, thirteen year old son, who performed heroic work in getting his sister out of the flaming rooms and wrapped blankets about both sister and mother, was unhurt, escaping almost miraculously. "I was standing in the kitchen door looking at sister starting the fire," he related. "It happened suddenly and I was knocked out into the kitchen. As soon as I could I dragged Frances out and mother ran out. Their clothes were on fire. I learned in school about using a blanket to put out fire so I rolled first mother then Frances in the blanket."

A painter at Joe Martin's, a quarter of a mile away, was the first neighbor to notice the smoke pouring from the Dean home. They rushed to the aid of the stricken family. Mr. Martin is a cousin of Mrs. Dean. Mrs. Dean and Frances had been removed when they arrived. All attempted to enter the house and get the baby out. Finally Mrs. Dean, despite her painful burns, rushed into the house through the kitchen, and carried the little son out the front door, all others having been turned back by smoke and burning drapes, etc.

The newspaper story – third & fourth column

Caryn continued.

George, Larry and I were sitting in our chairs and just chatting (Sunday), when I noticed Larry and I decided to get up at the same time. He walked to the front of the house (west side), and I walked to the side of the house the porch is in (south side). I felt the need to touch the foundation, and when I did, it felt as if it was full of love and a very happy place to be. When Larry walked near me, I invited him to do the same. He said he felt the love but also felt as if someone was enjoying his heartbeat.

So I decided to cover his hand with mine to see what would happen, and it felt as though there was a head grinding into our hands as if someone could not get close enough. My left hand felt the continuing feeling of love. We

stayed like that for quite awhile (less than three minutes) and then Larry pulled away.

Larry and I then felt a very cold spirit, which was perceived as a male. Larry asked me if I felt it too, which I replied, "yes," and he said there was a tobacco spat (from the male entity), which I was not aware of happening. Larry then felt as if someone was touching his face and remarking on how young he was to have a beard (actually, he had only a mustache). *At this point, he and I were voicing things for George's benefit.*

It was Margaret because she then moved to me and tried to 'enter me.' I kept reminding her she could not enter me or join me but had to stay at the house to take care of her family. After a few minutes of this I actually felt something I did not feel possible. What I felt was her hugging me (I think this the similar as an energy putting an arm around Gail's waist in Chapter 2).

Margaret Dean 1905

Caryn continued.

I did not think an entity could 'hug' someone, but I was being hugged, and she wanted me to promise her I would be back, and I will go back. She was so intense on staying with me that I handed her to Larry and he gently but firmly put her back on the foundation and told her good-bye and that yes, we would be back. She was satisfied, and nether her nor her family appeared again the rest of the evening.

Just when we thought our communications was over a 'spirit wind' came in (laterTuesday evening about an hour or so before we left about shortly after 2 a.m.). What happened is that the trees in the house foundation just above the rock that Larry had determined as belonging to the Native Americans rustled from the wind. The rest of the trees were still except these few. As the wind was swirling, I asked Larry what they wanted. He said that he did not know. I then said "Yes, we are concerned with Margaret and her family but we are aware and acknowledge that you are all resting here also."

The wind then tripled in velocity. It seemed to calm as quickly as it started.

The trees with the spirit wind

Caryn then finished describing observations.

The three of us sat and visited for another couple of hours in the middle of the property, and as we did, I actually 'saw' a figure in the back. I usually know an entity is there, but have never actually picked up on the physical features of them. This was a first for me, he seemed to just be watching us. Even though my first experience of an entity happened when I was ten, and I have never willed one to me, I now find that I am surrounded always by entities. It is almost like the TV show where she walks down the street and 'sees' all the spirits in their human form. I am constantly aware of spirit energies and being watched by them. This does not scare me but comforts me. I am puzzled as to if I am more sensitive or more welcomed by the other side.

Notes from Caryn's Experiences

I have several observations about Caryn's narrative about our our visit on June 25th and 27th. I want to make note of Caryn describing the front and back areas at The Farm and other notes.

1. When Caryn carried an entity to the 'front' and then to the 'back' area, it indicates to me that the 'peacemaker' entity she transported back and forth was possibly a former female person that wasn t an agitated male energy. Caryn's observation of the front and back talking in 'tongues' is an interesting thing that I can't explain. I am curious why the front and back entities behaved in this way. Were the 'peacemakers' in the back trying to smooth the dialogue with the settler family and us? It is hard to speculate without more information.

2. At the time Caryn stated that she and Larry stood up, both of them stood up at the same time. I found this event intrigingly interesting due to their spontaneity. There was no contrivance that I observed related to that moment by Caryn and Larry.

3. The crow taking off looked like it was startled, and the turn it took seemed like it was avoiding a predator or object in the air.

4. When we experienced the trees moving on Tuesday evening, I confirm that the night had no wind, and the cyclonic twister that swirled the trees in the house foundation did not affect the other trees on The Farm. The wind started abruptly in the seemingly

quiet atmosphere. There is no way to test this event, so it remains a matter of anecdotal conjecture.

Archives of the Family

Donald Dean

I sent this summary to Dan after our visit.

> *We three (Caryn, Larry and me) visited a neighbor to the south, Ray and Phyllis Harms, who said a Dean woman and daughter died there at the farm in a fire before 1945. We found that, at the cemetery on Homer Road, five Dean family members had died in 1928. Larry found a newspaper article that Margaret Dean (maiden name Doutre), her 16-year-old daughter, Frances, and four-year-old son, Donald, died from a kerosene explosion that started in the living room when Frances put too much on the fireplace on April 16, 1928. The explosion happened around 7 a.m., and the three of them died in the Saginaw hospital between 3-4 p.m. The house was not torn down, but apparently charred in the living room and*

was rebuilt/refurbished. Larry sensed that a person lost a limb near the northeast corner of the living room when they were rebuilding or otherwise working on the house.

Hazel Dean

Larry counted nine total foundations to the farm with a very strong oxen/bullock/cattle presence near the rear northeast corner of the barn. He felt a strong Native American presence between the farm and the creek. He does not feel a fire pyre presence, but a conduit may be nearby. The house foundation surrounds a red granite stone near the front of the house that was used by the natives for grinding various items. There was a strain on relations between the pioneers and the natives when the house foundation was built. Both Larry and Caryn felt a cold chill near the outside of the door entrance of the dairy shed attached to the south side of the barn where you apparently started your levitation with the milk pails.

When the three of us arrived Sunday, Caryn felt great confusion and disarray amongst the spirits. She assured them we were peaceful, so they quieted down. The starlings and crows were very talkative during our presence. On Tuesday night, just after Margaret stopped communicating with us, we were greeted by a spirit wind that rustled just the leaves of the trees that grew over the stone at the front of the house. It was a still night, and the wind did not disturb other trees, the grass or other objects around us.

We did find some other major things. I personally feel we have found the main entities connected with the house. The Native American influence seems relatively benign. More investigation on them, the servants, and some other yet unnamed or unidentified entities may be in the offing.

Front row: Francois (Frank) & Josephine (Serrurier) Doutre, Rosa (Doutre) Denome, Steven Doutre. Back row: Thomas Dean, Margaret (Doutre) Dean, Octave (Tab) Denome, Chester Legrow, Mary (Doutre) Legrow, Anna (DeGuise) Doutre, Edward Doutre. Far right: Francis Dean

Stirring Energies Beyond The Farm

On the first day, when Larry, Caryn and I visited The Farm on the 25th of June, Dan had a visit by an entity in Grayling, Michigan. He was in a gas station restroom when a female form came out of a stall and screamed at him. Here is his email describing what went on.

Hi George,
Since I sent you the email with the questions about the farm, things have started to happen. First several different lights blink out for a short time in my presence: my reading lamp, street lights, car headlights, yard lights at our cabin and at home, along with several other lights. I thought it somewhat odd but not sure what it was about.
Last Thursday I was in a men's room at a gas station in Grayling. I was washing my hands when a woman charged out of the men's room stall screaming twice as she ran at me with her arms outstretched and her hands open like claws. She made two or three steps towards me then disappeared. A little unnerving, I must admit. She was about 5' 7" and of medium build. She dressed like a gypsy with several bright colors and several layers of flowing garb. I could not make out her face other than I could tell she had a lighter complexion.
If you, Larry, Gail, or Caryn can help me understand what is going on, I would be more than happy to hear what you all think. Have we opened a can of worms? Is it a sign to back off? Mind my own business? I don't know, but I think she was PISSED! It was a definite attack. Any ideas anyone?
Thank you for your insight, Dan

Dan's experience was apparently triggered by our visit to The Farm. Our visiting The Farm 'stirred up' the energies. It is interesting that Dan's experience was most likely caused by our trip to The Farm at least 100 miles away. Of course, these incidents are anecdotal, but something caused Dan to experience his event.

I have seen a similar energy from the top of our shop west of Copper Harbor, Michigan crossing the road. I saw that it was female in form with flowing clothes like long rags that went along her arms and to her feet, but the clothes weren't in color, they were white. She had a hood on. Her size was approximately the same as Dan's description with a lighter complexion. Dan, Larry and me saw her at different times in the same outfit. Larry saw her from the shop roof crossing the road in front of our shop, and Dan and I saw her from our house outside of our window facing north. Their descriptions were similar to mine.

Caryn responded to Dan regarding the incident Dan had in the gas station restroom.

Dan,
I don't think it's as much that we opened a can of worms as that you are thinking about it, and I think the spirits sense that, so it makes you more sensitive. Does that make sense?
Since I was at The Farm last with George and Larry, I walk down the streets and see spirits all over. Some days it drives me crazy, especially when they taunt me or hang on me asking for something. I attribute it just to being more sensitive now. I have found that I can ignore them if I am not in the mood for it. Other days I smile and say, "have a good day."
I think the spirits know you respect them. I think they sense you are good person, Dan.
You can set boundaries with them, you know? That is what I did. I simply said in my mind like a telepathic message, "no one is allowed in my apartment," and they have respected me.
So tell them, "quit with the lights and making things fall. You guys are driving me crazy!" I am inclined to believe they will listen to you.
I too look forward to meeting you. I think you may find comfort when are energies come together in the same room.
Hope your week is good,
Caryn Ann

Dan responded to Caryn.

Caryn,
Thank you for your quick response. As you have noted, I have had a sensitivity for some time. My problem is, I never learned how to deal with it. I can see and hear things, but I have not tried to communicate since I was a child on The Farm. I have no idea how to talk to the spirits, and I am not sure I want to. It is just something I am not sure I want to bring into my life. Since I lived at The Farm, I have tried to put it behind me.
Since we started discussing The Farm, my sensitivity has grown stronger. I feel like I have enough going on in my life without getting into the spirit world. My biggest fear is to get into something I can't handle. I guess that happens with things we don't understand. I don't know what their limits

are nor do I know mine.

I ask about if the woman was sending me a message. I ask not out of fear, but out of respect for their peace. Am I causing them unrest? Are my actions causing them fear or harming them? I don't want to disturb them as I feel they have every right to a peaceful existence. Since I have no experience having a two-way contact with them, I am asking you guys to use your experience to help me understand. Perhaps they know what I am typing. As I typed the second paragraph, things have stopped dropping to the floor as they have been for the last few days. I hope so. I would like them to know there is no way I want to harm.

Dan

Dan asked if we would ask Margaret some questions regarding Dan's family events that occurred at The Farm. Here is his request.

George,

I would like any and all info you come up with for the old place. If you plan on going again, I would like to give you some questions about events and sighting from my childhood. Maybe Margaret could help me understand what was going on. Thanks for sending the last update. Wow, what a fruitful visit! I have no idea about what was going on around the milk house. Did Margaret say why she decided to levitate me? Can she do it to someone? You should try it! You might also ask her if it would be OK to take a picture of her and of the area. I think you would have to use regular 35mm film to catch the image. Just food for thought.

Dan

Here is the email from Dan asking those questions.

Hi George,

I finally got together a couple of questions for Caryn to ask at The Farm.

My dad would like to know what caused the milk box to fall the way it did.

My mom would like to know who it was that would stand at the side of their bed on several occasions.

Mom would also like to know who or what was in the storage area of the attic.

I would like to know who was making the bed shake and why.

I would also like to know who made the shadows both inside and out. No one was there, and there was no light source. Was there a reason for the shadows? For that matter, what was the reason for any of the events that happened there?

We would also like to know why all the other families were scared away from the place shortly after they moved in. How were they scared off? What did the forces there do to scare them away?

What happened to the last owners that made them destroy the house? Did they think that would stop the haunting of the area? Did it upset the spirits that the house was destroyed?

Does Margaret know who is there besides her family?

Can or do they travel beyond the confines of The Farm?

How would they feel about me visiting?

Would they allow pictures?

I still do not feel comfortable returning to The Farm. The answers to these questions may make me feel differently about a visit.

Thanks, Dan

Exchanging Perspectives and Encounters

I sent this message to the group.

Hi Dan, Caryn and Larry,

Dan, I agree with what Caryn has said. I still am pretty much an outsider looking in. I, as you know, have been observing what the three of you and Gail have experienced these past several months. As much as my eyes, ears and brain tell me, I am still a creature of my analytical habit. My general impression is that the apparent spirit world is as diverse as the living world that we currently reside in. Every spirit has its own needs and/or reasons for the diverse ways they behave. Apparently, they travel or exist in a multitude of levels that affect how they exist and have the ability to communicate with us. That being said, they are not all that different from how us earthly humans exist now. If that isn't a recipe for diverse happenings, I can't think of what else would be.

For some of us, they apparently are there for the viewing, as Caryn and Larry have been experiencing. For me, I just don't seem to be as priviliged for the viewing.

My gut feeling about your current latest experience(s) is this: Someone/something is exploiting your most vulnerable side with an irresistible urge/need to intimidate you. Being the kind and gentle soul you are, and given the sensitive accoutrements you possess, I feel you are letting yourself be used by these wayward spiritual influences. My fight or flee response to what you saw at the restroom would probably have been surprise coupled with intense adrenaline influences, extreme interest and a feeling that I would be pissed that I was being unwarrantedly accosted. Now, having all of those extreme influences of psychological and physiological affronts to my senses in such a short time is intimidating enough. As Caryn has said about seeing very many entities everywhere, I think you met what we can expect to see in our ordinary human lives. You are bound to run into an asshole now and then on a random basis. These random anomalies, whether in our world or theirs, are probably products of adverse conditions that affect their current or past history of behavior. Hence, we get to experience the end result.

At The Farm, I suspect there are one or more conflicting entities that just do not mix well with each other and/or human presences. So far, I agree with Caryn and Larry that we let them affect us only as far as we let them. Myself, I am in an apparent position of power where I can boldly go where you, Caryn and Larry have reservations. I just don't feel intimidated by what apparently is there and what we don't yet understand. An apparition or other overt influence could change that, but as Gail has reinforced about the two of us, we are positive people who expound their positive thoughts and influences to those around us. That is an advantage I/we have when dealing with this investigative situation.

George

Dan sent this message back to me.

Hi George.

I am back from Canada where I went to do some ATV riding and fishing. Well, I struck out fishing but got a lot of exploring of the wilds of northern Ontario. What a beautiful time to be there. The color is unbelievable. With our ATVs, we were able to get to areas that you would never get to any other way. We stayed at my uncle's fishing camp on a lake about two hours from the nearest anything. No power or plumbing, and the only heat was from an old wood cook stove like my grandmother used to cook on. Best of all, it is the most wonderful, peaceful lack of noise you never heard.

You know, George, I think you have a better handle on this spirit thing than perhaps the rest of us. You make a lot of sense to me. I think there are several levels of them. Yes, the spirit world must have jerks and a- - holes, just as we do in this world. Furthermore, it follows that there are weak and strong spirits just as there are weak and strong people. Because Larry and Caryn have been able to order some spirits to follow commands, it does not mean that there are not spirits that they cannot control. Heaven forbid that they should find out the hard way. Think about it. If the spirits of the house are afraid to go to some areas of The Farm, it must be that harm could come to them if they do. Could these areas have strong and/or evil spirits? I think if it is not good for good friendly spirits, it must not be good for good people who are spirit sensitive.

Please, Caryn and Larry be careful. I have chosen to not actively seek out spirits. If I come across them, that is fine, but I will not go looking for them, nor will I interfere in any way with their existence. As you know, I have had a life with many encounters with spirits. For some reason, when I have contact with them, it is not a big deal at the time.

After the fact, I start thinking about the event and why it happened. The lady in the bathroom was something new. Never have I had such an encounter where I could hear and see a spirit so vividly. I am sure by now they know that they will not be scared off, but I will question why it happened. I think I understand now that they wanted me to think about my relationship with them. I now I have a firm stance on how and when I interact with them.

I have decided that I will never go to The Farm again unless they need me for some reason.
I want them to live in peace as all of us would like to live.
Hope all is well, Dan

I then sent this message to the group.

Hi Dan.
Sounds like a great time in the true north. Silence is like a symphony in the right place and time. I hope the snow wasn't too deep and the coyotes and wolves didn't keep you up at night.
Well, almost every person familiar with the situation at The Farm seems to have an abundance of caution regarding something we have little information about. You, Caryn, Larry and the others who are on the periphery of knowledge about this ongoing investigation either stress caution or a reluctance to engage this unknown.
It is a healthy approach to an apparently possible detrimental 'situation.'
I continue to wonder what is there and why it behaves in the way it does. As of yet, I do not see significant danger to myself. I am wary, however, I feel there are many avenues open to me/us to understand. Being in fear of the unknown just doesn't wash with me, nor do I think it does with the rest of us. Winston Churchill's saying, "We have nothing to fear, but fear itself," seems appropriate for what apparently affects most of us. If it is a real threat, there should be a way to understand it without jeopardizing our well-being.
Perhaps, as in our world, these obtuse spirits could use some counseling or other intervention to return them to their 'norm.'
In my continually extending life, where I have had the benefit of education and experience wrapped around my Carl Saganistic base of understanding about the world around us, in addition to the experiences that you, Caryn, Larry and Gail have shared with me, I intend to combine an approach of our grassroots interests and efforts with some professional input and education. As time and resources allow, I will seek those latter avenues to help our understanding.
Care for my gut feeling of the moment? OK, I will tell you.

I see three of my closest friends/family who are adversly affected by something they don't understand. Certainly, I don't understand it either. Yet, I feel we are being bullied by the unknown. If it is warranted to have this fear of the unknown, so be it. It may need to be respected. But I personally want to know why I should respect it. Pure power of intimidation by spirits or people is an abuse of power.
But I also do not intend to be dead right in my pursuit of knowledge. However, I will listen to all of you if you think if I should pursue a different course.
Later my friends, George

Dan sent this message to me.

Hi George,
Thanks for keeping me abreast of what is going on. You have my position pegged. I would like to remain a passive observer of your investigation. I am very interested in whatever you find out. I also believe in keeping my promise to the 'Spirits of The Farm.' I would not want to put their peace at risk in any way. Things have returned to normal for me since the promise, and I would like to keep it that way. Not that it scares me, but if they go through all that trouble to get me to back off, then they must feel I am a threat in some way. I do not want to disturb their way of existence. They may have to be there for an eternity, and if I screw that up, it could turn their existence into a hell on earth. I would not want that. I don't even know what kind of threat I am to them, so I feel the best just as an observer without interaction. That way I feel the least risk of messing up the way they exist. With that said, all I can say is be gentle, be peaceful and be careful with your investigation of The Farm.
Your friend, Dan

I had a dream in vivid detail where several zombie-like people were staring straight ahead to the left of me. Also, I experienced a slap on my chest one night.

I sent this email to Dan.

The entities appear lost as many of us are on this side. It appears that a significant percentage of the personalities/energies over there (at least the ones we can see or sense) are on the lower edge of cognizance of even their own environment. Many don't even know they are dead. As I suspected, it appears many of them don't have the will, ability, intelligence or other wherewithal to move on to the greater plane.

One new thing I have found, among many other enlightenments, is that a personality needs to use a considerable amount of energy to make themselves apparent to us. Some of them are better at possessing or harnessing the required energy. They either have enough energy or they don't. I suspect who you saw was fairly inefficient at projecting themselves to you but had some sort of need to make themselves known for whatever reason. Also, I would doubt I would have seen the same energy that you did.

Although I have seen the same form of white zippies (my phrase) *at our place on the deck of the trailer, I don't think I have seen anything else like what you have seen at your Aunt's.*

This being said, I think about what you, Caryn and Larry have seen in years past. Larry has seen distinct Native American visuals (among others) and Caryn has seen many other personalities. Caryn has seen literally multitudes of personalities when she has been driving. From the information I have been reading, it appears there are roughly just as many personalities on the other side as there are over here. I am somewhat surprised that we don't interact much more with them than what is reported.

I suspect that ignorance of them from societal, cultural, religious and other circumstantial influences is combined with a general mismatch of sensitivities of our side and their side. I am beginning to believe that there just is not a strong enough concerted need by a significant majority of us on each side to increase our awareness of each other. Also, on average, I think the majority of the ones we do usually make contact with are by accident and that neither of us care or have a significant need to pursue our contacts with each other. In fact, which probably is no surprise to you, Gail, Larry or Caryn, I wonder to myself, why is it that the four of you don't find it irresistible to find out what makes it possible for all of you to communicate as you can? If I had any of your abilities, I would be testing, searching and chronicling the gifts the four of you have. Your gifts entice my interests to pursue further what there is to find.

So, I think your ability collided with a low-level personality that couldn't express itself more fully, liked to play tricks, or both.

As you may recall, several months ago, I dreamed/envisioned a group of personalities that were dressed in 1800s clothes. One of them had a large nose and made it quite evident that I should pay attention to him. After about a couple minutes, they dispersed. Well, a couple nights ago, that large-nosed personality made his appearance to me again. He wanted a fight, so I was in the mood and we tangled. I was so pissed at him for being pissed at me. We tangled and I kicked his ass. He left abruptly. A while later, that same night, another personality and I had the same altercation for the same apparent reason. The same result happened. I did not recognize who the second personality was. When I woke that morning, I realized the resemblance of the two personalities, but not the intent of why we had to fist fight. If I had clearheadedness, I certainly would not have had a fight with him, but try to find what the motive was for it to behave the way it did.

Dan replied.

As for your encounter, I find this very interesting. You are now sharing the world the rest of us have known for some time. Welcome. Could you detect what they were looking at? Do you have any idea why they came to you? I wonder if they are here to help you, or if you were there to help them. It raises many good questions. Do you think one of them might be the one who slapped you on your chest? You are closing the gap between the two worlds. Keep me posted on any new events.

Sounds like these spirits are good if you sensed no bad vibes. Don't take this as a firm rule, but I think, for the most part, it is true. I think I would rather meet the spirits you did than the screaming gypsy I met. Did you try to talk to them? If so did you get a response? I have to wonder what you feel after being shown this different world. I know you are curious, but you are a realist. Where does this event leave you know? It can be a gift or a curse. I hope it is

a gift to you.
Take care my friend, Dan

I sent this message to Dan and the group regarding his interpretations of the events. The first part refers to that dream where I saw several figures looking to my left. They were dressed in 1800s attire, and they showed no emotion. They weren't moving about. Also, in this message I philosophize about energy detection.

Hi Dan,
Here are my responses to your insightful questions with my tangental memories.
I felt they were not looking at, nor seeing, anything. They were zombie-like. As I recall, most were staring straight ahead to the left of me. In fact, this 'dream' sequence I had was more of a motion picture feeling where I wasn't interacting with what they were doing. Many other dreams are like real life where I am living and experiencing my dream. This one wasn't interactive. If they were there to help me, they sure were doing it in a subtle way that I didn't detect. I didn't see if I was helping them. However, this was an apparent first where I was able to see them in distinct form. For all the chest slapping, shoulder touching and neck caressing I have been physically experiencing these past two years from these apparent entities, in addition to the rapid heart pounding when they get very close, I would have expected, in retrospect, that they would have made more of an effort to communicate with me directly. You, Caryn and Larry have had substantial visual encounters while you were wide awake and in daylight. Mine are surrounded by circumstances where my guard is down during sleep periods. I remain skeptical as to whether I truly am experiencing sensations that indicate they are from the other side.
As far as whether one of them may be the one who slapped me in the chest, none of them had made any indication of that, nor do I have any reason to believe one of them did it.
I did not think to try to talk to them, probably because they did not seem to be a chatty group. Plus I had only about 15 to 30 seconds of visual contact with them. During that time, I studied the first big-nosed entity for over half the

duration, then I gazed at the others, noticing they were generally the same dress and divided about equally between men and women. None of them had the inordinately large nose that the first one had.

I sense a trend here that resembles a Catch 22. The more I scrutinize these situations, the harder, or less likely, the entities seem to be able to, or want to, contact me. On the other hand, if I 'make' myself believe there is something/someone there trying to contact me, it appears to be a self-fulfilling prophecy that I will be contacted. It is akin to something like this. The most skeptical of investigational skeptics seem to never, in my remembrance, find credible evidence or personal experiences to change their minds. Conversly, those who are apparently the most gifted in experiencing these contacts don't/can't/won't delve into the why of what makes these experiences tick.

Have you heard of a sensitive person with a history of multiple and continual contact with the other side express a keen interest with investigational abilities and credentials to try and find out why and how this phenomenon apparently goes on? Even though we may be talking about different levels of existence outside of the modern sense of reality on Earth and beyond, if it exists, which it appears to do, there is an explanation for how and why it works. On top of all of this, I have a significant amount of respect and suspicion of the human brain. Deep down, how can we ever truly trust our mass of chemically charged neurons and synapses to keep us on the level where we can function in the best interest for ourselves and others? I have found there is an exception to every rule, law and certainty in virtually every instance of our reality. That includes each one of us.

With that being said, who can we trust if we can't truly define what our brains are telling us about ourselves, others or other things? That is part of the reason that I am naturally comfortable with being skeptical. It makes the most sense to me, and I am fond of simple common sense because fact is stranger than fiction when you dig deep enough.

Alas! If my (deceased) dad came to me in broad daylight during waking hours and slapped me in the chest in front of you, Larry, Caryn, Gail, Michael Shurmer (world renowned skeptic) and Sylvia Earl (world renowned medium), and all of you said, in effect, "George, you just had a crossover experience!" I would need it repeated under controlled circum-

stances at a later date with the appropriate measured controls and verified by other research experts before I would begin to accept that something significant happened.

However, remember, there is a crucial and significant difference between something being believed it happened versus knowing that it happened. That is what I am looking for in this situation we call The Farm (and related happenings).

What do you think, seriously? All replies are welcomed.

Caryn replied to me.

My senses to the happenings of the spirits that surround us always have been heightened since I started on The Farm. The place has lots of energy, both positive and negative. I am enjoying my new sensitivities and have learned to even call spirits to me now.

When my mother-in-law died, I was in the room, and the spirit within her contacted me. Her eyes turned black, and I knew I was exchanging with it. Over a year ago, her spirit screamed to me it was trapped in the vessel. About two days before she died, I asked her spirit to wait for my sister-in-law to arrive from out of state. I knew she couldn't handle it if her mother died before she arrived.

The third encounter was just a brief moment before she died. My sister-in-law called me in a panic as her mother's breathing was labored. I jumped in the car for the 22 mile ride, and when I arrived, the spirit immediately looked at me through the black eyes and asked if it was time. I answered, "yes," and, in a split second, she gasped a breath as her body tightened for the passing, and her vessel was gone. That was back in early June (2007), and I still think about the energies I felt.

I believe that I brought someone home from The Farm the last time, and they are still present, but quietly 'peek' at me. I am just letting them learn to trust.

I am careful, Dan but understand the only danger is if I allow a spirit to enter me, although my vessel may have room for two. It really does create a

lot of problems. I am strong and will not allow that to happen.
Caryn

Reminiscing after those Visits

In November of 2023, Caryn sent me a message providing more detail about her experience from those two visits.

> *The farm seemed to be divided into three areas. The house area where the family was, the back barn area which seemed to be foreign workers, and the Native American burial grounds. The back barn area seemed happy, the workers were content. They always seemed to be at peace and having a good time, even inviting me to play games. Larry, who is a visionary, was able to 'see' them and what their purpose was in life.*
> *The native burial grounds were forbidden to me. The first time I attempted to go back there, the birds warned me not to, so I did not, out of respect. The next trip, I knew where the line was distinctively. I crossed it and immediately was in fear. I knew then it was burial grounds because they did appear to me in form and told me I had no business bringing my Christian beliefs to them. Out of respect, I left.*
> *Years later, I went to The Farm again with a friend and met George and Larry. I went into the barn in the burial grounds area and found what I believed to be their portal* (actually, Larry provided the direction to the portal). *I stepped in it and felt as if I was being sucked down under ground. I yelled at my friend and the guys to grab me. They did and pulled me out of there. I was very thankful they listened to me right away. I will never return to that area again as they really do not want me there.*
> *George and I went to the Homer cemetery a few days later to look up the graves of the family. While there, we ran into Grandpa (Frank Doutre) who was bewildered as to where his family was. I told him they were trapped, and he could move on without them. He wanted to wait for them to cross over.*

Frank & Josephine Doutre 1924

Caryn continued.

On another trip, I reasoned with Grandma and Mom, and once I told them Grandpa was waiting for them, they agreed to go into the light. I have not been back to The Farm since, as I feel my job there is done, and I am not willing to intrude as an unwanted guest in the burial grounds.

Back to The Farm, upon my first visit, I came to realize that there had been a family that had a bad house fire and that some members perished. They made their presence known to me. I described them to George and Larry in detail. The next time I saw Larry, he brought me a newspaper article that described everything I had said with great accuracy.

Finally, validation.

Larry at the stone used by Native Americans as a tool in the house foundation

I messaged Larry asking him to explain a comment where he said the stone at The Farm was of interesting to archeologists. Here is his reply.

> *There were two archeologists researching the Saginaw Indians and trails/roads in the Saginaw area. I ran across their research when I was researching Council Rock.* (a book he is writing about a Native American shaman meeting place in north east Gratiot County along the south side of the Pine River) *Fred Dustin is one of them. He and this other fellow were a University of Michigan professor and student who wrote to each other and explained their findings.*

Six

Interview with Hazel Buzard and Lynne Hurkley

I arranged an interview with Hazel (Dean) Buzard and her daughter, Lynne (Haddix) Hurkley, on August 11, 2007. Hazel was the one sleeping upstairs when a fire occurred on April 16, 1928.

Lynne contacted me in July of 2007 after I posted in an Ancestry.com forum asking if anyone had information about the Dean and Doutre families in Mt. Haley Township, Midland County in Michigan. The interview was at Lynne's place in Midland, Michigan. Here is my recap of the interview I sent to our investigation group.

> *My meeting with Lynne and Hazel was interesting and informative. It was intriguing to have Hazel recount what happened on that tragic day in 1928. Here are the main points we covered. I recorded our one hour and 45 minute interview. Both Hazel and Lynne are level-headed and sharp as tacks.*

Lynne & Hazel

1. Hazel was about 5-6 years old when she was upstairs with her sister, Emma, who was asleep. Around 7 a.m., Hazel looked down the stairs, saw flames and woke her older sister, Emma. Emma quickly grabbed Hazel and dropped her from the east porch roof into the snow bank where Anna Mae followed. As Hazel read the newspaper article Larry found about thetragedy, she said "That isn't right," about a couple points. (However, she didn't elaborate.)

Left to right: unknown, unknown, Thomas & Margaret Dean, Anna Doutre with baby Hazel Dean, Frank & Josephine Doutre. The window above the porch is where Emma dropped Hazel to safety.

2. Lynne has about 15-20 pictures showing Deans and Doutres. One picture shows Frank Doutre on the back porch with family wearing a white shirt and black pants, similar to Larry's vision. There are 7-10 pictures that show the house from all angles. Pretty cool. I will scan and send all pictures to you when I receive them. The copy center was closed today, so Lynne couldn't make copies for me, but will mail them later.

3. Hazel has had zero spirit connections at The Farm or any other time in her life, according to her recollection. She essentially said in the interview that she doesn't believe in spiritual contact, but my impression is that she is open-minded. Hazel comes off as no nonsense.

4. There are/were several other buildings on The Farm that show up in the pictures that we haven't seen on our visits. There was a large chicken coop building immediately northeast of the house where Frank and Josephine moved into after the tragedy.

5. I asked Hazel about servant quarters near the barn. She said they couldn't afford servants and didn't know about any.

6. At night, Hazel said she was afraid to use the outhouse.

7. I told the story of the person who lost their arm near the bay window. Hazel nor Lynne ever heard of that.

8. There is a picture of the north side of the barn showing livestock where

Larry sensed them to be.

9. There is a picture showing the north side of the house where an orchard stood, now a cultivated field.

10. Lynne brought out a tightly rolled picture, the original of the four-year-old who is pictured on the front page of the 1928 newspaper headline story.

11. There is a separate (wedding?) picture of Margaret as a beautiful young woman. Her husband, Thomas Dean, is in an accompanying separate picture. They appear to be in their late teens or early twenties.

This interview didn't reveal any new details about the fire at the house. However, Lynne's pictures were a wonderful addition to the story. Her pictures of the people and the buildings add a new level of depth of where they lived.

Donald & Hazel Dean

Lynne told the story of the fire in her own words.

*So, as my Mom (*Hazel Dean Haddix Buzard) *tells it, the men went out to the barn in the morning and her Mom* (Margaret), *and sister* (Frankie), *were getting breakfast. Frankie went to start the stove (she usually would go out and get a cup of kerosene and pour it on the coals) but she was blown back across the room. Margaret was burned because she was in the kitchen and Donald (almost 4) got scared and hid behind the stove.*

Mom and her sister, Anna Mae, were upstairs in bed. When they realized that the kitchen was on fire, her sister dropped her into a snow bank from the upstairs window and jumped after her. She said that her family was visiting the Martins, about 3/4 of a mile down the road the night before the fire. People said that they saw a truck stop by their house that night. She said that her dad had put in two tanks up by the road: one for gasoline and the other for kerosene. Apparently, they had filled the kerosene tank with gasoline, and that was the cause of the fire.

Front row: unknown, unknown, unknown, unknown, Anna Dean, Hazel Dean, unknown, unknown. Middle row: unknown, unknown, unknown, Frances Dean, Norman Dean, Rilla Dean. Back Row: Elsie Dean, unknown, unknown, Jim Dean, unknown.

The Dean children June 1950. Jim, Hazel, Anna, Rilla, Elsie, Norm.

Seven

The Sixth Visit to The Farm

The Year: 2008

I wanted to follow up with Hazel and Lynne's suggestion to visit The Farm, so I sent Lynne an email to coordinate their meeting there.

Hi Lynne,
Myself, Caryn and Larry plan on visiting The Farm this Sunday, August 10th at 1 p.m. If you and/or your mother would like to join us, you are welcome. Primarily, Caryn promised Margaret that she would return after our last visit. I think it would be interesting if your mother paid a visit there, so that Margaret may do some communicating with Caryn regarding your mother's presence. Also, since you are Margaret's granddaughter, there may some interesting communication regarding you. There is a possibility that there may be nothing happening between Margaret and you and/or your mother. While we are there, there is nothing in particular that you or your mother need to do. Caryn and Larry can carry on conversations with us seamlessly as they communicate in their own way with Margaret and others.
Thanks, George

Lynne replied to me.

> *George,*
> *Thanks for thinking of us, but I'm very sorry to say that, on June 14th, my mother passed away. For March and the first part of April, my mother was getting increasingly weak, and she told me just to get her home. She knew she didn't have much time left. I brought her back to Michigan in the middle of April and set it up for caregivers to live with her in her home. I stayed with her a couple of weeks, returned to Mesa, and then returned to Michigan about four weeks later to stay again. I was on the way to Michigan for the third time when I got the news that she just couldn't hang on any longer. She took a nap and died peacefully in her sleep. I'm back in Arizona again. Good luck going to the property. Please let me know what happens. Maybe Caryn will find Mom along the way.*
> *Lynne*

This chance to involve Hazel and Lynne with a visit to The Farm, in retrospect, is a missed opportunity to explore the interaction of Hazel with her mother Margaret. Perhaps no interaction would have occurred. We probably will not know because this opportunity wasn't apparently meant to be.

Hazel's Obituary

> Hazel Haddix Buzard of Midland and Florence, Arizona went to her Lord Saturday, June 14, 2008 after a short illness.
> She was born in 1922, in Mount Haley Township Midland County. The daughter of the late Thomas and Margaret (Doutre) Deane, she was one eight children, all of them deceased. She and her husband Otto O. Haddix; whom she married Oct. 14 1944, started Reliable Plumbing and Heating in 1949. After Otto's death on June 22, 1984, she married Victor Buzard 1992. Victor proceeded her in death on June 22, 2002. She enjoyed sailing her catamaran. She was always there for her children and grandchildren in all their activities, including: 4H, figure skating, girl

scouts and boy scouts. In her later years she was much involved with quilt making and supplied all her grandchildren and great-grandchildren with beautiful quilts.

A women of many talents, mostly self-taught, she had a zest for life that she passed on to her whole family. Hazel made friends everywhere she traveled. She will be greatly missed by many grandchildren, nieces, nephews, friends and neighbors, both in Midland and Arizona.

She is survived by her children James (Nancy) Haddix of Ann Arbor, Lee (Susan) Haddix of Midland, Lynne (Woody) Hurkley of Mesa Ariz., and Lori Haddix of Midland; grandchildren, Leigh, Sam and Ben Haddix, Erin (Haddix) Broestl and Christofer Haddix; as well great-grandchildren, Moira, Henry and Bronwyn Broestl. Hazel is also survived by her step-children, Barbara (Chuck) Wallace of Apopka, Fla., Victor Jr. (Linda) Buzard of Raleigh N.C. Janet Buzard of Ysilanti, MI and Peggy (Duane) Schwab of Midland.

The Energy Runs Deep

Despite the absence of Lynne and Hazel, this visit was a very interesting day of discovery. Caryn witnessed Margaret ascend as she crossed over. Also, with Caryn and Larry's sensitivities, we discovered a Native American portal. The following summary by Caryn captures the essence of what we discovered. There is more detail in Caryn's summarization about Margaret when she sent an email to Lynne, later in Caryn's summary.

Caryn's summary

When I arrived to the farm for my last visit, I was anxious to get to the task at hand, so I immediately went to the foundation of the house and sat down. Margaret came to meas soon as I sat down and chewed me out for taking too long to come back. At first, I thought, "Wow, I wasn't aware she knew the concept of time!" I soon realized she was so agitated because the rest of her family had moved on. She explained to me she was ready to go, and just like that, she was gone. This is the first time I literally watched a spirit cross over. It was fascinating to watch her fade away into the light. I felt a little sad, as I knew I would be missing her in the days that followed. Finally, my work felt complete with the family.

Next, I decided to go to one of the side buildings and see what communications I could achieve. I felt the same sense that I was not welcome in that area of The Farm.

Standing by the willow tree, I felt the same sense of peace and joy that I had always felt in that area. I walked back to the big barn and leaned against the corner of it for help in discovering its mysteries. I was told to go to the cornerstone. I asked Larry where it was, so he did his thing and directed me to the cornerstone. I went to that area of the barn and felt a strong negative force. There was a distinctive dip in the ground in that area (the northeast corner in the interior) *of the barn. I felt as I was being pulled down toward it, and it took every ounce of everything I had to keep from being sucked in. I noticed Larry felt at peace there. My friend walked right through it with no indication that anything was going on. I know now I am not welcome in that area of the farm ever and I will never try and go back.*

Larry, my friend and I walked to the opposite corner of the barn (the southeast). *Outside we felt an overwhelming sense of peace. This is where our Mediterranean friends reside. They are most welcoming and I felt as if I could stay there forever.*

I do not feel any need to go to the farm anymore as I feel my work is complete. This whole experience, from the onset a few years ago, has validated what I have done.

Caryn sent this email to Lynne the same day (August 10), telling her about her grandmother Margaret passing to the next level.

Hello Lynne,

I feel the visit to the farm today was successful. I promised Margaret I would be back.

She immediately came to me today as I sat on the foundation of the house. She was clearly 'ticked off' at me for not coming earlier. She was waiting for me, but the rest of her family had moved on. After I gained her confidence, I encouraged her to go. In my urgings, I informed her that her daughter had

passed on. She questioned me, "My baby?"
I said "Yes, your baby Hazel." She then continued by asking where she was.
I replied, "I don't know, but my guess is with the rest of your family."
With reluctance, but a sense of peace, she told me good-bye. I watched her go up into the heavens. (In the east at a 45 degree angle). *I feel good knowing she is with the rest of her family. The rest of the story concerning the other spirits occupying The Farm, we will share with you later. I just wanted you to know about your grandmother right away.*
Luv, peace and blessing to you always, Caryn Ann

I followed the next day with an email to Dan recounting our experience from my perspective.

Hi Dan,
Well, very interesting again, and a lot of details to recall.
As you probably know, Larry, myself, Caryn (and her friend) visited The Farm this last Sunday at about 1:20pm. We had some very interesting experiences and discoveries.
First off, Caryn made contact with Margaret after some effort. (I will be recapping her description of events according to her email to Hazel's daughter, Lynne Hurkley, which you should have received a copy). When Caryn was able to contact Margaret after some difficulty, Margaret conveyed some unhappiness that Caryn took so long in getting back to The Farm. (This appears curiously new and significant that Margaret is showing a sense of elapsed time).
Caryn had expressed to me, about two months ago, that she had an urgency conveyed to her that Margaret wanted Caryn to return to The Farm for a visit (as Caryn said she would last year when we last visited The Farm). Margaret 'told' Caryn (during this visit) that the whole rest of the family had moved on and that she was the only one of the family that was left. Caryn then told Margaret that Hazel had died recently. Margaret asked where Hazel was now, and Caryn said she didn't know, but that she probably was with the rest of her family.
Shortly afterwards, Caryn observed Margaret's energy form ascend into the

eastern sky at about a 45 degree angle as it/she passed out of Caryn's view. After that, Caryn had no presence felt for Margaret being at The Farm. Shortly after this happened, Caryn told me what had just happened and gestured towards the direction that she saw Margaret leave.

About an hour later, Caryn tried to re-contact Margaret near the back porch of the house but had no contact with her or any other members of the Dean family. Caryn has a good feeling of closure with Margaret and her family. After Caryn told me what she had communicated with Margaret, she said she still had a strong feeling to not go beyond the willow tree that stands about 20 feet east of the outhouse, as she had felt on previous visits.

After Caryn said that, I continued following the north property line (where the field started) photographing the outhouse and three buildings going to the east. As I walked beyond the third building, I saw Caryn standing at the entrance to the barn, which surprised me, given the previous hinderance she received minutes before. After I approached her, she told Larry and me that she was getting a message to find the cornerstone of the barn. We needed some more guidance, so Larry asked for some. (During this time Larry spread his arms to each side facing to the northeast corner of the barn to receive the energy guidance.) He was instructed to proceed to the far northeast interior corner of the barn.

As Caryn and Larry went there ahead of me, Caryn said she felt a strong vortex feeling in the center of the room nearest to the northeast interior corner of the barn. She had to steady herself against a pillar, because the power of the vortex made her dizzy and unsteady. She remained there for a minute or so until she regained her equilibrium. She then told us that this is the location of the portal as described by the Native American in my spirit session February of 2006 (as reported in Chapter 4).

Caryn and Larry were both quite leery of the power and unknown nature of what they had just found. Larry then confirmed to us that he also felt what Caryn had found. At first, both of them did not want to get any closer to the portal of energy. During this time, I continued to take pictures from all angles. After they made evident their unease to approach the portal location, I instinctively stepped directly into the spot where they both said the center of it was. I felt nothing especially significant when I did this. Larry and Caryn eyed me expectedly, waiting to see what might happen, which nothing did.

Caryn and Larry with the portal at the tire

During the entire time I spent in the portal, I could feel the effects of my adrenaline levels being heightened, perspiration and general sense of 'presence' in a noticeably higher state, similar to a sense of tension of my being in the moment. It is a hard feeling to define. My time standing in the portal location was, in essence, a non-event as far as experiencing energies, compared to what Caryn and Larry felt. My description of how I felt may have been a reaction to the newness of our group's dynamics from experiencing such a new energy, at least in part.

If this event happened now (2026), I probably would be able to observe what I was feeling with more clarity related to energy sensitivities due to my increased understanding of energies. I recognize that my ability to identify energies is not as sensitive as Caryn's, Larry's or Dan's. Now I have the ability to recognize that some energies are present.

Back to my email to Dan.

After I stood in the portal's center, Larry and Caryn took their turn. Both of them felt quite a significant power of an apparent tornado motion of many energies. Larry said he felt much death from the Native Americans who died from smallpox. Apparently all the energies in the vortex are Native American related. He also said that the entities/energies at this location are very old, nearly ancient.

Just before Caryn stepped in, she put her hands against the portal's vortex and said she could feel the wall of energy. Just before I stepped in, Larry said the vortex had a diameter of about 8 feet. As he stood in the center of the vortex, he added that it extended 40 feet into the ground and 400 feet into the air. After Caryn stepped out of the vortex, she said that she felt that the energies communicated she still was not welcome.

Larry, on the other hand, as he stood in the center, said (to this effect), '"Wow, this is really cool to feel this!" Larry said that the entities did not care one way or the other whether I stood at the center of the vortex.

Larry sensing direction to the portal

After we three had our stand in the vortex, I continued to take pictures from all angles. Then Larry told Caryn that he was instructed to direct Caryn towards the silo outside to the south of the east door to the barn.

We all converged to the outside southeast corner of the barn where Larry said, "This is the location where the dark-skinned people/workers were lo-

cated." Caryn and Larry stated that this location was very peaceful. Larry said that this is the location that the man on the back porch looked at us with distrust and suspicion. This entire corner on the outside of the barn had a concrete foundation that was intact and disappeared under the grass and soil as it went away from the barn's corner.

After this, we worked our way back to the house. Larry said that he also went through the door of the gambrel-roofed building where Gail had the arm put around her waist when she and I visited there previously (our visit in February 2006 in Chapter 3). *As he went into the doorway opening and back out, he felt a strong heat, similar to a fire.*

As we summarized events, Larry said he saw a bald man with a bowler hat in a brown suit, with a flattened nose and a pot belly that pushed open his suit coat, who was scurrying around the house area. He appeared to be a banker or law enforcement individual, like a revenuer, since there have been indications of a still on the farm. He also said that there was still a strong Native American presence around the stone in the house foundation. A final comment Larry made was that there are many entities still about The Farm.

What do you think of that Dan? Pretty interesting, huh? Of course, many questions arise to me as far as what you remember about the locations where we had these experiences. What was in the northeast corner of the barn when you lived there? How about the outside southeast corner of the barn? Anything else?

I will probably be sending the 50 or so pictures I took after I return from up north in late August. Some of them are quite cool.

Catch you later, George

Larry & Caryn with portal lower left. This the only picture where orbs appear. The noose at the right of the picture is an artifact, assumably. It doesn't appear in any other pictures. Also the dark liquid-like in the image is probably an oil residue or other liquid.

After I sent Dan the email recapping what we discovered on our sixth visit, he replied with the following email the next day.

Good to hear from you. As I recall, the whole north end of the barn was the milling area. There was a large mill set up to grind corn for feed for the cattle. As I remember, it was powered by a spinning drive belt wheel on the side of the tractor that engaged when the PTO (power take off) *was activated. That turned a long drive belt about 10 inches wide and about 25 feet long in a loop. You would put the belt on the mill, then line the tractor up and put the belt on the drive wheel. Then back up until the belt was tight. You had to get everything just right or the belt would fly off which was very hazardous! To me, the mill seemed like a big growling monster. I knew it could rip off an arm in a heartbeat. It could also suck you into its grinder and spit you out before you knew what happened. I have no idea if anyone was ever harmed or killed by it. With all the whirling, whining and growling going on with it, I can imagine that thing itself as a vortex to hell.*

On the southeast end of the barn was the exit to the barnyard for the cattle. The whole south end (of the barn) *was a milking parlor. On the southwest end was a huge water-filled cooler where we stored the milk in 10-gallon milk cans. These would be picked up by whichever dairy we chose to sell the milk to. The southwest corner is where my levitation started. I will have to talk to Dad to get you more details.*

Just a bit of a side note, the lights going off and on for me has started again. Oddly enough, it started a couple of days ago. Kind of matches up with your visit to the farm. I am not sure what role I play in this whole thing. I kind of thought that when I promised to leave the spirits alone it was with Margaret and her family. Perhaps the deal was with someone else or the whole group of spirits there. I am not sure why I am a factor to them. I mean them no harm and that should be clear to them. I have to wonder if I hold a key or have a power that is a threat to them. If so, I have no idea what it is. Any thoughts?

Thanks for the update, Dan

Larry, Caryn's friend and Caryn in the north east part of the barn.
The portal is at the tire location

I followed up with an email to Dan.

Dan,

I think you are unusually 'gifted' with your sensitivity, whether you want it

or not. You can treat your gift like diabetes, either take medication for the rest of your life (constantly, more or less, demanding the entities leave you alone) or do nothing and let it affect the rest of your life. I think overall, The Farm and its attached entities are not specifically going out of their way to harass you. I think you are an easy target for the entities that surround us all, due to your unwanted gift. Apparently a bad gift.
George

Larry, Caryn & Caryn's friend on the north side of the barn. The portal is on the left side of the picture.

Dan responded.

Hi George,
I do not see the activity as harassment from the spirits. I don't see it as a curse. I feel it is just their way of getting in touch with me. I do feel it a blessing to have any connection with the spirit world. Sometimes it is a good thing and other times not so good. Kind of like everyday life, isn't it? I just find it a bit hard that they can't get their full message to me. I would love to know what they are trying to tell me. I am not good at guessing at what it is they are trying to tell me by the signs they have sent my way. I know I will forever have a connection with The Farm. I just feel like I am one step from hearing

them clearly and understanding what they are trying to say.
The problem is that I have no idea what that one step is or in what direction. Bottom line is, I am here to help them if they need it, but they need to make it perfectly clear what they need from me and how to do it. They need to use words I clearly understand or visions that are perfectly clear in meaning and content. If I cannot be of help to them, then I feel we should both leave each other to our own existence. Just a flash here, I wonder if they are trying to help me somehow. I never thought about that. I hope they show me either way.
Dan

I contacted Caryn the next January, hoping to gain insight from her views during our last visit to The Farm.

Hi Caryn,
I am thinking about your recollections of last August 10 at The Farm. Your own words are important to what we have found there, and I sure would like to have them, so we have them for posterity. The more detail, the better. When I compile enough of this information, I intend to put it into some form of media, whether it be a book, summary or other form. I have a good amount of pictures from my photos, the old newspaper, Lynne Hurkley's family photos and some probable illustrations that I or others may create and contribute.
You are the major reason why I have been able to progress in this investigation, since you were the first one to start providing answers and leads that have led us to where we are now, which is a better understanding of what was and is going on there. Of course, we have seemed to establish that Margaret's family stayed there ever since about 1928 before she was the last one to leave last August 10.
Also, apparently the portal that you and Larry teamed up to find was a spot where the Native Americans burned or otherwise disposed of the bodies of those who died of smallpox there. In my research, I have found that, most likely, there is a type of rock or rock formation under the portal location where there is a pronounced rift or crack in the rock where a portal is

normally found. Also, the Ojibway or Sauk Indians there usually built their medicine (sweat) lodge(s) over these spots where their shamans practiced their craft of providing the community with their spiritual needs. I would bet dimes to donuts that an archaeological dig there would turn up some interesting telltale clues to confirm this.

I have been reading R. Guiley's Encyclopedia of Ghosts and Spirits *which has been quite enlightening. As I expected, and I stated in earlier email summaries, the spirit world is exceedingly disparate, scattered, elusive, non-conforming, fractal and hard to understand.*

Dan asked a while ago, "Why they don't they just come out and make themselves clear?" Well, apparent to what I have been seeing in my research, a typical majority of spirit events we have been able to attribute to The Farm experiences and sensations are not the product of intelligence, per se, but of energies that are not usually completely random, but not of an order (if any) that lends itself to explanation to us.

So, Caryn, I am looking forward to your first-hand account of what chronology of events you experienced there last August. I welcome anybody's reflections of what you have experienced then or since also. In fact, I really would like to hear what you have to offer.

Then I received this email from Dan about energy events happening to him.

Hi George,

I was sitting at a computer in my aunt's office. It has a 3' x 4' window that looks out onto the porch. It has a roof over the entrance and a wall with two large glass windows across from the one I was looking out of. I will draw you the layout when I see you.

To start with, I saw a ball of light about .5 inch in size do a one-and-a-half turn spiral from the bottom up. Later, I saw a flash of light about 12 inches in size on the outside of the outer wall of windows. The spiral was on the inside of that wall. Then I saw the lamp that was on a bench behind me come on. It was a circular florescent lamp on a movable arm. The odd thing here is that I saw the reflection clearly in the window I was looking out of but did not notice any added light illuminate the room. Another thing is,

when I turned to the bench, I traced the power cord to its unplugged end in plain view on the bench with no electrical outlet near it. Then another spiral in exactly the same size, shape and placement happened outside.
In each case, I was seated at the computer in the same chair but at different angles to the window. I did go outside to investigate the outside lights and found nothing and nobody that might have caused the lights. I didn't get any vibes good or bad from any of this but that is not uncommon for me. Just thought this might be of interest to you. I am hoping that you may have some insight on these happenings.
Best to you my friend, Dan

Caryn replied to Dan and the rest of us.

Dan and others,
I believe what you are experiencing is your 'guardian angels' (if you will) that are there for you; they are probably feeling it is time to make themselves known to you. I prefer to refer to them as Spiritual Guides. I personally have five. Three have made themselves known to me over the years. I also refer to them as 'soulmates.' They are the spirits that are destined to be with me forever. One is my Nana (my father's mother). I felt the connection when she was alive and now know she is by my side always. She is my source of logic and reason when dealing with others. I have a 'cowboy' guide who reminds me to laugh at life when it gets me down. I have a 'sumo' wrestler guide who toughens me up when the going gets rough. The other two have not made themselves known other than I know they are there.
I went to a tea leaf reader once in Ontario while visiting my friend Gary, and she was the one to inform me of the five. After that, the second two introduced themselves (already knew Nana was there), and the other two, in ways you described, let me know they were there.
I believe when our 'vessels' die, our spirits move on to complete the job they started to do. Hence, why there are so many spirits hanging around, they don't have their assignment yet or they aren't ready to give up their job and 'go into the light' or some, I believe, just want to harass us.
My cowboy is from my frontier life I have discovered as a past life. I worked

with a woman about twelve years ago that was in that life with me as my 'partner.' We think she was the guy and I was the lady. She is also one of my soulmates as is my granddaughter, Olivia. She likes to sit on my lap with her heart touching mine cuz she says it feels good. I remember the same feeling when I was a girl with my Nana (yes, my grandkids call me Nana). The last time I saw my Nana still in her vessel was at my 12th birthday party, and I remember sitting on her lap to get that feeling, and her spirit 'telling me' it was going to move on but not to worry, she would always be with me. Her vessel died two days later peacefully watching TV in her favorite chair. She is still with me, as promised.

I suspect I will make the same promise to Olivia one day. Dan, do not fear. Relax and accept and let them be known. Perhaps they feel you are going to need them or do because of the other ways you seem to have been harassed (the flickering lights and such). They probably want you to know they are there for you. Take comfort in that.

George to address the curiousity or, like you said, 'irresistible' urge to figure out what's going on, I do not have to totally know what is going on. I embrace it, understand it and feel blessed to have the gift and use it when necessary. You know how I handle the spirits. I let them come to me at will and kick them out if they are bothering me too much. I have a fun spirit that lives here in the rental with me, even my friend gets a kick out of him cuz he is constantly playing tricks on us.

My mother-in-law's spirit (her vessel died two years ago this coming June) came to me the other day and asked me if her husband was okay now and could she leave? I assured her he had adjusted fine since her death, and yes, she could move on. I saw her go into the light. She was my second. My first was Margaret the last day we were at the farm. I love my gift and sensitivity, but I have had to learn to be tough and firm with the spirits when they try to interfere with me too much. That is against my nature, but so I can stay sane, (okay as sane as I get) I follow through with it.

Dan, always remember that you are in control. I made that mistake for too many years, letting them walk all over me. I could tell you countless visits since I was nine years old. It wasn't until the last time I laid on my Nana's

chest that I understood what was going on.
Luvs always, Caryn Ann

Dan replied.

Hi Caryn and George,
Thank you for your insights on my recent events. I found these interesting because it seems a step farther than I have been before. I don't feel like a participant, but more like an observer in these events. I feel like I am blessed enough to see them but still don't understand them or how to interact with them. This is where I wish I had some of the skill you have, Caryn.
George, perhaps I can help you understand my lack of aggressiveness in going deeper into this side of existence. I find my experiences interesting, but I lack knowledge and skill to go farther than what just happens to happen. I know there is a whole other world there. As in our world, there is good and evil. I feel it foolhardy to go there without the knowledge of how to do it safely.
The other factor is that, just as in this life, I have no need or desire to harm or generally screw up anyone else's existence. Easy to do when you don't know what you are doing. I feel no threat nor harm from the other side, and I want them to know they should feel the same of me. I guess harmony is all I seek. I do wish I knew enough to help them if that is what they seek of me. We will all get to the other side soon enough.
At that point it will become more clear to us. As of now, I am in no great rush to get there. I feel my life here is so full and good that I want to live it to the fullest while I have the chance. I am blessed with great friends and family, and the time here with them will suddenly be too short. That seems to be the case when all of the sudden, the years have slipped by. I want to spend my time with the here and now, the rest will come naturally. Right now I am engaged in making this life as good as it can be for those treasured friends and family. That is how I want to spend my time here. I will take the time I have for the other side when I get there. I still find all of the events of great interest. I always seek knowledge, so I will ask questions. I guess what I am trying to tell you, George, is that I highly value my time here. Lots of things

interest me, but I will not let them consume me. I am very busy living this life. It has twice been nearly done and I am so blessed to still be here. I am going to make the best of what I have left here. There is much love, laughter and adventure here in this life. I seek all of it I can while I can. The best part of life is sharing these gifts.
Dan

I replied to Dan.

Dan,
Specifically, from what I have been reading lately, your experience is a combination of variable personality energies with your apparent heightened ability to observe and/or experience them. As I suspected from our earlier experiences and investigations, I am finding out that personalities on the other side are just as lost as many of us are on this side. It appears that a significant percentage of the personalities/energies over there (at least the ones we can see or sense) are on the lower edge of cognizance of their own environment. Many don't even know they are dead compared to their physical body. As I suspected, it appears many of them don't have the will, ability, intelligence or other wherewithal to move on to the greater plane.
One new thing I have found, among many other enlightenments, is that a personality needs to use a considerable amount of energy to make themselves apparent to us on this side. Some of them are better at possessing or harnessing the required energy. They either have enough energy or they don't. I suspect who you saw was fairly inefficient at projecting themselves to you but have some sort of need to make themselves known for whatever reason. Also, I would doubt I would have seen the same energy that you did.
Although Gail, you and I have seen the same form of 'white zippies' (My phrase. I saw white commas about four inches tall, when I returned from Lansing) *at our place on the deck of the trailer west of Copper Harbor, I don't think I have seen anything else like what you have seen at your Aunt's.*

A depiction of the energy Larry, Dan and I saw. She wore a hood. We saw her at our place west of Copper Harbor during our investigation at The Farm. She was in white clothes.

I continued.

> *This being said, I think about what you, Caryn and Larry have seen in years past. Larry has seen distinct Native American visuals (among many others) and Caryn has seen/sensed many other personalities. Caryn has seen/sensed literally multitudes of personalities when she has been driving. From other information I have been reading, it appears there are roughly just as many personalities over on the other side as there are over here. I am somewhat surprised that we don't interact much more with them than what is reported. I suspect that ignorance of them from societal, cultural, religious and other circumstantial influences is combined with a general mismatch of sensitivities of our side and their side.*
>
> *I am beginning to believe that there just is not a strong enough concerted need by a significant majority of us on each side to increase our awareness of each other. Also, on average, I think the majority of the ones we do usually make*

contact with are by accident and that neither of us care or have a significant need to pursue our contacts with each other. In fact, which probably is no surprise to you, Gail, Larry or Caryn, I wonder to myself: why is it that the four of you don't find it irresistible to find out what makes it possible for all of you to communicate as you can? If I had any of your abilities, I would be testing, searching and chronicling the gifts that the four of you have. Your gifts entice my interests to pursue farther what there is to find. So, I think your ability collided with a low level personality that couldn't express itself more fully, liked to play tricks or both.
George

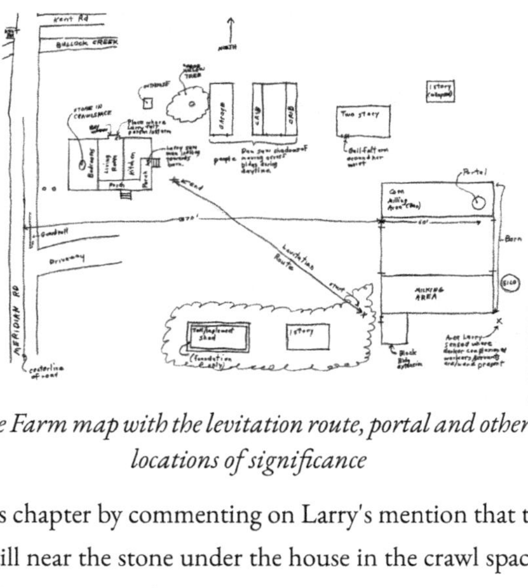

The Farm map with the levitation route, portal and other locations of significance

I want to end this chapter by commenting on Larry's mention that the Native American presence was still near the stone under the house in the crawl space. If that was the case, I suspect that there was a conflict of energies between the Native American energies and the family energies. I think this relates to Caryn's transporting the different entities back and forth between the front and back of the boundary approximately 20 feet east of the willow tree. Caryn's communication with the crows is apparently explained by this explanation.

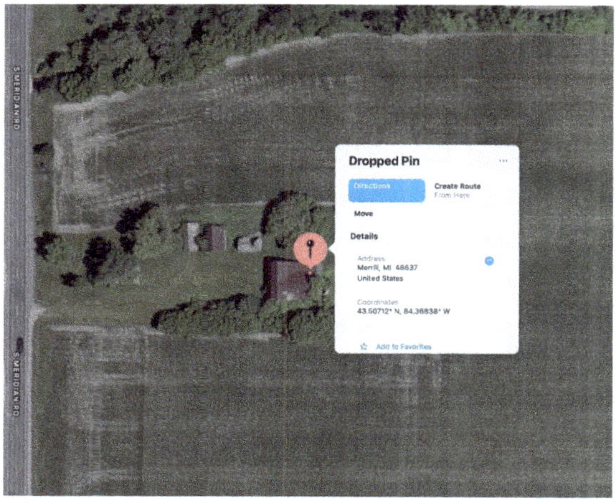

The Farm map with the levitation route, portal and other locations of significance

Eight

My Out-Of-Body Experience Dream Related to The Farm

I had a dream April 23, 2009 that had several similarities to an Out-of-Body Experience (OBE). While I doubt the significance of this dream, it is interesting how it relates to The Farm events. The responses to my experience illuminates OBE events that others have experienced. I shared this dream with Dan, Caryn, Larry and Gail's and my friend Peg Kauppi. I think there are clues in all of our responses, and in my dream, that relate to events at The Farm.

Within the day that I had this dream, I made two sketches of what I had seen. I had a vivid memory of those scenes. I dreamed in color, and I marveled in the realism of my sense of distance while we flew together and when I floated in the room. I have had flying dreams all of my life with various forms of controlled trajectories. I have experienced uncontrolled flights and flights where I was in control of my direction and speed. When I fly in my dreams, I experienced a euphoric thrill due to the inherent knowledge that this is not normal in my usual world. I never have felt in danger during my flights.

Here's my letter to Peg.

Hi Peg,

Last night I had a first-time experience while dreaming that Gail suggested I share with you. I am not sure the significance of what it may mean, but it was quite fun and vivid. I am sharing this email with my brother, Larry and my friends, Caryn and Dan because they will identify with several parts of it, especially the first part.

I can tell the story now in a way that I have inserted some of my recognition and interpretations of what the scenes apparently meant.

I was walking north on Meridian Road (The Prime Meridian of Michigan's original surveyors plat) in southwest Midland County as I approached within a quarter mile of The Farm, which you know about, that Dan lived on back in the 50s and where Larry, Caryn and I have been investigating for the past few years. (Our latest visit last August, we finally located the portal).

At this point in the dream, I realized my destination was to meet Gail at The Farm, but at that moment when I saw The Farm about a quarter mile in the distance, Gail joined me, and we were diverted by an unseen power to continue our walk immediately to the west. As she and I trekked west, we came to a plain house on the side of the road near the coast of Lake Michigan. Gail suggested we stop there and seek advice on how to return to The Farm. Inside the house was a plainly furnished typical Midwestern farmhouse and a man who seemed to know Gail. He suggested we stay for a while and "acquire the needed ability" to take the next path for our trek.

In the next couple of hours, we met with other people, apparently in a holding pattern of their apparent trek. Most of them seemed to have a 'clingy and sexual need' type of personality. Also during this time, I had to resist an uncontrollable strong urge to fly around the different rooms while we visited with others in the house. The man we originally met told us that we needed to put in our time to continue our journey back to our destination, assumedly, The Farm.

After a couple hours (in real time of my dream) of mingling with these other 'people,' Gail gave me the hairy eyeball, and it was understood that it was time to leave. So, I floated down and sat next to her, put her left hand in my left hand as I put my right arm around her (I was behind her). We then lifted and floated up through the roof to survey the landscape.

As we hovered arm in arm about 100 feet above the ground, we were looking northward across a green valley about 100 yards across with a 100-yard tall hill on the east and west side of the grassy valley, which had gently sloping sides. The valley continued northeast and southwest.

Surrounding us in the valley were about 100 people as individuals and couples who were all floating around as if it were a Sunday afternoon picnic on a perfect spring day. I got the impression these entities had also recently 'discovered' their surroundings, much like a bird just learning to fly from its nest. I told Gail to look behind us because the valley opened up to Lake Michigan with a city of skyscrapers, similar to Chicago, a few miles away on the shoreline (the shoreline nearest us). The scale suggested to me that we were about a mile or two above the lake level. We turned back to face north and watched the others serenely glide around the valley dressed in white flowing garments with some tinged with pastel trim. The only time the garments fluttered was when some of them were gliding. Others remained motionless, apparently taking in the view. The day was clear and vivid and apparently in clear daylight, although there was no glare like a full sunlit day now would be.

After a few minutes, I said to Gail, "That's the way we need to go." She used her hand to gesture toward the northeast. So we started gliding about a hundred feet off the ground north through the valley avoiding trees as we made our way. A while later that afternoon, we were crossing flat farmland, and the colors were drab, if at all visible. It was apparent we needed to call it a day to rest.

We appeared in a room of a place I was familiar with and asked Gail if she knew where we were. She said yes, and we settled in for the night.

From what I have read and heard others describe, it appears this was an astral projection (or OBE). I have not had this type of experience before, but I realize there is probably much to be interpreted and learned from this. If this is what the near other side looks and feels like, it is one cool place that I look forward to experiencing. I welcome your take on this, Peg, and anyone else as well. Thanks, George

Peg replied.

Wow, George, this rocks!
I think the Meridian Road is especially pertinent to you because you know the origin. As in the Prime Path, you projected onto the other side or maybe the meridian, meaning middle, as in you were straddling the middle between earth plane and the other side. The Farm, is maybe symbolic of the 'goal:' the ultimate end of the trip or heaven. And you and Gail are diverted as you are finding out more about your life plan together. (Of course the guy knew Gail, everyone does!)
You might have thought of the clingy people as more earthbound people as sexuality is often thought of as an earthly event. And being that you were more inspired to experience your new 'tricks,' might infer more of a spiritual mindset having to suppress the urge to fly all over the rooms. (It's really great that you realize that a lot of stuff can be learned here, and you might even think of where you are suppressing urges in every day life.)
There's so much here, George. It's wonderful. The only other thing that really strikes me is that you used the number 100 five times, and I don't know if you feel there are different vibrations to numbers, but the importance of that number is: "Keep your thoughts totally aligned with God's love and light." Your positive thoughts and actions make a huge difference especially right now. I actually think I'm also supposed to pay attention to that message from you, so thanks! I am green with envy, what an experience!
Peg

Larry replied next.

Hi George,
Your dream sounds like astral projection to me, but the astral projection I'm familiar with is done when you are awake while putting yourself into a 'trance-like' condition that is very close to the sleeping/pre-dream state. I suppose the phenomenon is possible during dream state, and it's great that you can remember it.
The common elements to your experience with the astral projection are: the 'floating above and seeing,' called teleportation, the existence of a house with a gatekeeper/guide/advisor and the association with others having the same

simultaneous experience. When we engaged in astral projection (when I was at Hillsdale), the fellows doing it at the same time were visible and interacted with each other during the teleportations.

The part of your astral projection that doesn't seem to be present is the concept of layers, or planes of consciousness within the environment of the 'dream.'

What I learned was that there are dangerous 'spirits' in the higher planes and one who is not experienced with the practice of astral projection can be taken advantage of by them.

Your ultimate defense, if you encounter an evil confrontational spirit, is to look the entity in the eyes, place your forked fingers (index and middle on your left hand) on each side of your nose against your face and point over your left shoulder in a 'get out of here!' motion.

You might have noticed that there is no talking during the teleportations, except for the gatekeeper who keeps on warning and implicitly preparing you (I also witnessed this fact separately). Expressions and visual signs are the methods of communications.

A man from the Bahamas, Ace Stewart, was a classmate of mine at Hillsdale, and he enlightened me on the subject. (He is now Dr. Ace Stewart, PHD in philosophy and writes source books on these subjects). He said that a lot of the 'black magic' in the Islands involves astral projection. He showed me the 'cast out' symbol when he found me on the third or fourth plane way before I was experienced enough to be there. He said people who go into comas as a result of voodoo spells may have been put upon by these evil 'spirits.'

I was in a house that had at least four floors, and sections of the top two floors were blocked off and unattainable. But I found a way to get in and got frightened out of my gourd until Ace showed up and saved me.

Larry

Dan replied next about my original email to Peg.

Hi George,

What can I say!!! This is awesome! I have only had one such experience. I would love to have another. I think you were blessed to have this and to have Peg to help you figure it out. I think it is way cool that Gail was with you in

this experience. Now that is bonding! Has she experienced any part of this dream herself? Thanks for sharing this with me.
Dan

I replied to Dan.

Dan,
Larry had an enlightening reply. I will forward it to you. He continues to surprise me with experiences I was unaware of. It just goes to show, apparently, how far behind or inexperienced I am compared to others. Actually, we all have our own level we seek, much like water does when affected by gravity, I think.
Gail didn't directly share in this dream, but we have shared directly in others several times. Quite cool how we can wake in the morning and share common experiences. Actually we just had a common direct shared experience. We had ice cream with strawberries, bananas and chocolate sauce. Now that was some shared experience. Wish you could have shared it with us. Here's to your new experiences.
Cheers! George

I sent this reply to Larry.

Larry,
It was quite easy to remember. In fact, it was very vivid and not hard to remember, so I am thankful for that. Also, it seems interesting that the portal seems to be a power source or gateway for everyone, unless I was diverted by them as I approached The Farm. I have been assuming, for no really firm reason, that it is/was a Native American based phenomenon, but it makes sense now that it shouldn't necessarily be that way.
Also, apparently I just visited the happy novice layer during my visit. Guess the time is not ready for me to venture the next steps.
You keep surprising me with your past experiences. I googled Ace Stewart, but haven't found him yet.

How good to hear about the Council Rock project. (Council Rock is a book Larry is working on in northeast Gratiot County, Michigan near where our Quidort ancestors settled in the middle 1800s at a location known to have Native American energies) *Can't wait. Since that spot was such a center of communication, it seems likely that some form of portal or other energy field may have been in the area. Doesn't it seem like there may be more around there?*

We could probably keep finding things from that area for years.

Thanks for getting back. George

I sent this email to Peg after she responded to my original email.

Hi Peg,

Thank you for the insight. Your independent views touch on aspects that I would not have thought of. The Prime Meridian of Michigan was mentioned by Larry a couple years ago, but I had forgotten that he mentioned it was a crossing of an important thoroughfare for the 'white' man and the nearby creek (Bullock Creek, which is the name for the high school Gail, myself, Dan and Larry graduated from about 15 miles away). That creek was a thoroughfare for the Native Americans before the 'white' man. That seems to be an interesting confluence of energies.

A lot of the events I recounted, and what you explained, seem to have a strong base in symbolism, which to the uneducated and inexperienced, often seem to create a confusing staccato of vibrant and rapidly occurring experiences that don't make sense. I also see where it is difficult, at least for now, for many people to categorize what happens to them in their Saganistic view of the world, as they commonly know it.

Luckily, Sagan advises to keep an open mind to all the evidence (before testing it, which in the case of what I have previously related, is difficult, if not impossible to do right now).

A curious aspect to what you noted is the number 100, where I referred to the number of yards, feet and people. When I originally wrote my description, I had no reason (in my conscious reasoning as I look back) as to why I would be concerned with that number as compared to the subject of yards, feet

and people. Interesting to me, but probably self evident to others as to the significance. Also, your reference to the quote and its relation to the numbers is intriguingly (I'm somewhat at a loss for words) flattering to me. Overall, I like the direction this experience seems to be taking me. It seems almost too good to be true, given the history of mankind's preoccupation with the 'regular' world we live in at the present and have in the past.
George

Dan replied about Larry's response.

Hi George,
Wow this note from Larry is very interesting! I always thought the astral projection happened during sleep. I thought I only had one experience like that. I don't remember the whole experience, but I do know of a building like Larry described. I have been to it or one like it. I always thought it was a dream, but it seemed more real than a dream. Now I think I have had two experiences. I wonder what this building is? Does it have a name? I don't know how I got there or how I got back. I have always wanted to be able to do this on demand. I do know it can be dangerous, so I would only want to do it with a guide. I remember a guide on my first experience, but I don't remember one on the trip to the building. We will have to discuss this further next time we get together.
Dan

Caryn replied, however I don't have a record of me stating that she had three experiences.

Yes, George,
You are correct in that I have had three experiences. My first experience was about 15 years ago after I had surgery. I did quit breathing three times, and I do remember that. The third time they were getting the paddles out as my heart had stopped. I was able to start it again on my own. This was when I realized my spirit was looking down at my body. I suddenly wanted to live.

Before this realization, I had visited another plane. I had seen 'different levels of faith.'

It is hard to explain, but this is the closest I can get. There were different rooms in the heavens that had people with different levels of faith in them. It was made very clear to me that this is the reason we have different religions. Everyone was content where they were. No, I did not see a light, but I did know I was in that better place for a few moments of time. I walked away from it feeling comfortable in my faith and not feeling that 'Catholic guilt' because I don't attend church anymore.

The second was about 13 years ago. I was having a problem with someone ordering parts under my name for the company I worked for, and I could not figure out who it was. One night I felt a compelling urge to 'leave my body,' so I did. I had to go from my home in Whittemore to Oscoda, Michigan. There was one stop on the way. It was in Tawas, Michigan. My sister lives there, and she was having trouble with her newborn waking up screaming a few times a night for no apparent reason. I remember hovering over her crib and hearing the train, which shook their house, and I realized that was what was making her scream. I informed my sister a few days later. I did not tell her how I knew. After that, I arrived in Oscoda above my fax machine (it was on my desk), just in time to see who was ordering the parts. Luckily for me, my supervisor (the two of us had discovered earlier we were life partners in a past frontier life) was open to my method, so her and I figured out a 'tangible to the rest of the world' way to nab him. The trip ended with me zipping back to my body, and I remember waking with a start to the alarm that was set for my husband.

My third experience was when I lived in Lansing a few years ago, and a friend of mine was cleaning my chakras and lining up my energy fields as I had an important test the next day. He said he could feel me leave the chakra closest to my groin area. He knew I was hovering, and he was careful to leave my portal open while I did what I had to do, so that I could enter back in my body when I was finished. I do not remember leaving, but I remember hovering over as we wondered what was going on. My spirit felt scared, and I remember feeling an intense pain as I went back into my body.

These along with my communcations with the other side have helped me to be more comfortable with who I am. I have a jokester spirit living with me in this rental, and, in the neighbor's yard behind me, is the biggest, ugliest, gnarliest looking pine tree which houses spirits. When I get home late at night, I frequently see a body running toward the tree. When I go near it during the day, I feel the same 'get outta here' feeling I had when I was in the back part of The Farm. I try and go near it about once a week to see if I can get friendly, but usually I leave quickly as I don't like the negative vibes I receive.

I drew two sketches of my astral projections.

I made the first sketch the same day I had the dream/Out of Body Experience. It shows the relative positions of us and others as Gail and I hovered about 100 feet above the ground. This first perceptive shows our position with other people and landmarks. We traveled along the westward road from our location about a quarter mile south of The Farm. The city was about 10 miles in the distance. The single circles are individual people and the double circles are couples.

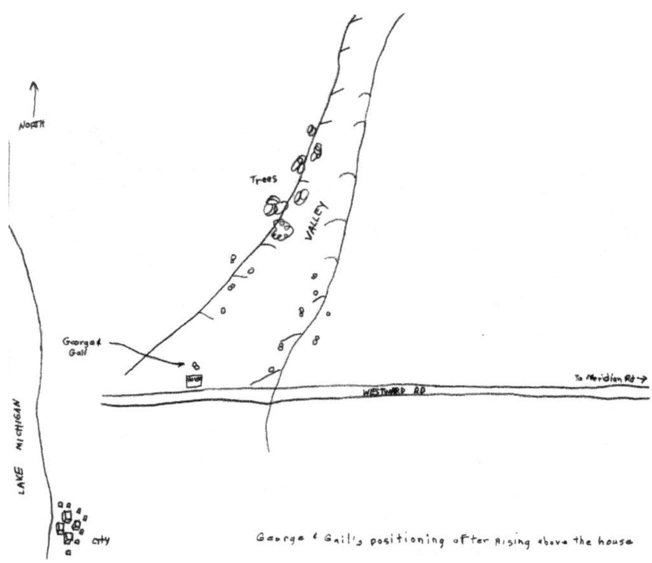

The second sketch shows our perspective in more detail: the single people and couples with a more detailed perspective of the valley hills and trees.

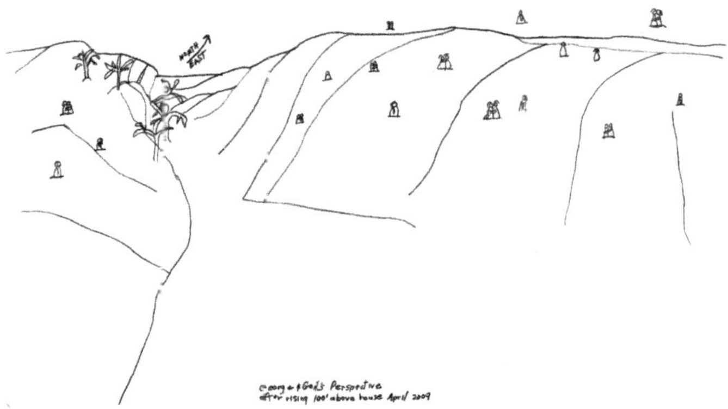

This perspective is from Gail and I when we hovered above the house. The people are about 100 feet above the ground as we were. This view was in color with green grass, green trees with a blue sky.

Nine

The Prime Meridian

The Meridian Road running north and south on the west side of The Farm began as the Prime Meridian of Michigan when the state was surveyed in the 1800s. I started researching the Prime Meridian after Larry told me of a surveyor being killed while surveying near The Farm area. Although I didn't find the information I was hoping for regarding the Native American residents, it was a learning experience nonetheless.

I sent the following email to the Survey Museum in Lansing, Michigan regarding that surveyor Larry mentioned.

> *I am doing research regarding a person or party who surveyed Michigan's Meridian in the early 1800s in the southwest part of Midland County, Michigan. We have found a reference that a surveyor was killed by the local Native Americans due to a possible conflict with nearness to their sacred grounds. This person could have been named Porter. I have found that this survey was done while Michigan had a governor named Porter. Also, the nearby township is named Porter.*
>
> *Our main interest is this person or party who interacted with the Native Americans. Apparently, military troops were dispatched from Saginaw, Michigan to assist in completing the survey.*
>
> *Are you aware of this person or events related to this part of Michigan's history? Do you have any people who we may want to contact for further investigation?*
>
> *My brother and I are compiling historical facts from this place and era that may relate to several curious sets of events we have become aware of starting*

as early as the mid 1950s to the present day. My brother and I were raised in the Midland County area.
Thank you, George C. Bailey

I received this email from Mr. Ingram, who was listed as the curator of the Survey Museum.

I'm going to copy a couple of friends in MI (He was located in Virginia) *who may be able to help with this. I suspect the person you will really want to be in touch with is Norm Caldwell, but I'm not sure if my email for him is still current. One of the guys I copy will be able to help find him.*
Regards, David Lee Ingram

Soon after, I received this email from Mr. Caldwell.

Mr. Bailey,
Your inquiry of 4 October 2010 has been forwarded to me for a response. The Michigan Meridian in southern Midland County was surveyed in 1830 by Robert Clark Jr. This Deputy Surveyor also laid out the Town and Range Lines in the region. The subdivisional survey of Porter Township, T 13 N, R 1 W, was originally contracted to Charles W. Christmas. Mr. Christmas found the conditions in the field so adverse that he left the region and, on 19 February, 1831, wrote a letter advising the U. S. Surveyor General that he had gone to Pontiac. Snow on the level high ground was 24" deep and over waist deep in the swamps and low ground with the temperatures reaching as low as minus 30 degrees.
Mr. Christmas is alleged to have left two of his men in camp at half pay while awaiting the spring breakup. They both eventually left the camp and returned to 'civilization.'
Mr. Christmas never returned to the region. (He later became one of the founders of Minneapolis, Minnesota, Hennepin County Surveyor, etc.)
In 1852, Austin Burt was contracted to perform the subdivision of Porter Township.

I replied to Mr. Caldwell.

> *Hello Mr. Caldwell,*
> *We will try to establish the source of our information regarding the purported death of the pioneer surveyor and forward it to you and others listed in this email. Having some myths disproved can be as valuable as verifying others. At this time, we will continue digging for information regarding our ongoing investigation. We are focusing our work on and near Section 28 of Mt. Haley Township, just east of Porter Township. From our research at the Clarke Historical Library in Mt. Pleasant, Gratiot County (Michigan) Historical Society in Ithaca (Michigan), Grace Dow Library in Midland (Michigan) and other sources, we have determined that there was an apparently significantly-sized Native American settlement and thoroughfare that existed near the northwest corner of that section where the meridian was laid out. The introduction of the settlers in the early and later 1800s seems to have contributed to the mixture of several facets of conflicting cultures.*
> *Thank you for your help, George C. Bailey*

I asked Larry what source he used regarding the email.

> *George,*
> *Bruce W. Carlson published a book named* Midland in 1973, *published by Pendell Publishing Company. On pages 5-6 in the first Chapter, named 'Early History' he wrote: "... In 1819 the federal government purchased a large tract of land, including the area to become Midland County, from the Indians. Known as the Treaty of Saginaw, this agreement allowed the government to immediately begin surveying its newly acquired lands. The Indians, however, soon began to fear the survey on with his strange equipment which he used to divide up the land. Hostility towards the surveyors increased as they moved westward throughout the Saginaw Valley. While contented at first to express their anger verbally, the Indians eventually took direct action by killing a surveyor, named Porter, who had been working the western areas. As a result of this incident, the government established a*

military post at Saginaw to prevent any further outbreaks of violence."
Larry

I received this email from Mr. Caldwell.

We have a complete listing of all field crew personnel for these U. S. Deputy Surveyors. There is no listing for a Mr. Porter. There is a similar 'legend' regarding a person of the same name who is alleged to have died 'in the service' in northern Newaygo County, Michigan. Several searches (and research) have been made of the site with no supportive evidence being recovered.
We do suspect that these 'legends' are derived from the eventual death of Robert Clark Jr. after he crossed the Grand River near Grand Rapids, Michigan, in Town 8 N, R 11 W, on 20 March 1837.
To our knowledge, Robert Clark Jr. was the only U. S. Deputy Surveyor to die while on the job during the original government surveys in Michigan. One other U. S. Deputy, who resided in Michigan, died while on the job, but, that was near Sheboygan, Wisconsin.
The Deputies were not required to list their camp persons (cooks, packers, etc) on their contracts or field notes, so there is a remote possibility that one of these support persons could have succumbed, and there would not be a record of the event.
I trust you will find this information of value, and should you desire additional research, please feel free to contact me at the above email address.
Further, in your continued research, if you are able to document supporting evidence of the 'legend,' we would appreciate being advised.
If you are researching the surveyors activity in the region, we would suggest obtaining a copy of our book Surveyors of the Public Lands in Michigan, 1808-2000, *of which we have several remaining in stock. Purchase price is $50 plus $5 S&H, total = $55.00.*
This publication contains a complete biographical sketch of each U.S. Deputy Surveyor that worked in Michigan, his family, employees, contract locations, memorandums and indexing.
Another book of interest would be the Special Instructions to Deputy Sur-

> veyors in Michigan, *available from the M.S.P.S. Institute, 220 S. Museum Drive, Lansing, MI, 48933. (formerly the Michigan Museum of Surveying)*
> *Norman C. Caldwell, P.S. Shiawassee County Surveyor*

I also received further update from Mr. Caldwell.

> *Mr. Bailey,*
> *Mount Haley Township, T13 N, R 1 E (*Where The Farm is located), *was subdivided by Robinson Thomas, D.S, in 1832.* (It was probably about this time that smallpox was exposed to the Native Americans at The Farm). *The subdividing surveyors usually made note of any Native American encampments, sugar bush, trails or significant features. If you have not done so, we would suggest a visit to the Archives of the State of Michigan Library, second floor, east wing, in Lansing where they have microfilm copies of these field notes.*

After my research on Michigan's Prime Meridian, I sent this email to Dan, Caryn, Larry and Gail, summarizing my findings. Although I didn't find significant information about the Native American presence at The Farm site, I did find that the Meridian is important to the state and the people who settled there.

> *Hello Fellow Interested Parties and others,*
> *Last Thursday, I went on a research binge here in Lansing to see what I could see about The Farm's location near Michigan's Prime Meridian.*
> *Originally, the meridian survey was contracted to surveyor Charles Christmas, who started it about January of 1830 but left camp due to two feet of snow and minus 30 degree temperatures. He returned to Pontiac but left two assistants in the wild, telling them that he would return when the weather broke. However, he never did, and the assistants eventually gave up and returned to civilization.*
> *Charles Christmas then went to survey the original plat of Minneapolis where a lake, street and statue or plaque are named after him. However,*

Charles Christmas said he screwed up by plotting the original first streets of Minneapolis to parallel the Mississippi River and not on a north south grid as it should have been (a lot like Midland first streets).

In 1831, the job was awarded to Robert Clark Jr, a son or other direct relative to the Lewis and Clark fame. He completed the survey and very likely is the first white man to encounter the inhabitants of The Farm's location. I found this information at the Michigan Museum of Surveying on Museum Drive, just south of Michigan Ave.

In 1832, Robinson Thomas surveyed Section 28, Mt. Haley Township where The Farm is located. I went to the State of Michigan Archives to observe his field notes. Unfortunately, all that he mentions is "Crossing a small stream..." as he laid out the east/west section line over Bullock Creek. I then asked about any field notes that Robert Clark may have made when he laid out the meridian a year earlier. The assistant said that there is no record of any notes, but there is a file of maps with possible notes regarding the base line survey and meridian survey. As I looked through this file of original documents and maps, there was nothing about the meridian. However, there were several original maps of the original baseline which at that time was the original state line that ran from the southern tip of Lake Michigan straight east to a point just south of Toledo. There was also a notebook/document dated 1816 from the state surveyor of Ohio, a Mr. Tiffin (who was later governor of Ohio and had a town and county named after him) that instructed the surveyor of the state border/baseline to "take the utmost care that your survey line be accurate because all other surveys in the territory of Michigan would be dependent on it."

In a book at the Surveyors Museum, I noted that there was "a concern that the state boundary was being carried out by an Ohio man." The beginnings of the so-called Michigan-Ohio war were being sown already. At a later date, the state baseline was moved north to its present location that is the county line between Ingham and Jackson County and lines up with Eight Mile Road, which is the northern boundary of Detroit.

At about the same time, the Upper Peninsula was included as a part of the Michigan Territory as a reward for giving Ohio the Toledo Strip, where the original baseline/state border survey line was relocated a few miles to the north. It was very cool to have these original maps and documents in my hands. There seemed to be a lot of energy in their midst.

After looking through these files and maps, the assistant suggested I visit the State Archaeologist on the fifth floor, which I did. Unfortunately, when I dropped in on Mr. Anderson, he was prepping for a meeting that was scheduled shortly, which a fellow office worker brusquely reminded him about. However, Mr. Anderson was very gracious and seemed quite interested in what I had been researching in the five minutes I used to review my facts. He agreed that the Clarke Library in Mt. Pleasant, Gratiot County Historical Society in Ithaca and Dow Library in Midland were the best spots to do my research. He said that if I had any questions, I could email him. He gave me his business card. He also said that he gets information and requests for various aspects of Michigan history all the time, just like mine.

So, I found this quite interesting about events leading up to The Farm. Michigan's Prime Meridian (Meridian Road) was selected because Michigan's oldest town was/is Sault Ste. Marie, and they wanted the Prime Meridian to go straight south from that town.

Ten

Lynne Hurkley Obituary

It was quite a shock when I learned of Lynne's passing. Of course, she died much too early. She was instrumental in providing a great amount of detail about our story. She was the daughter of one of the main people that experienced the tragic fire in 1928. Through her, we have many pictures that enrich our story. Lynne contacted me after I posted in an Ancestry.com forum asking if anyone had information about a family in southwest Midland County, where I said that some curious things happened in the recent past. Lynne was quite interested in our story and was enthusiastic to share her information.

> ***Lynne A. Hurkley***
>
> *Lynne A. (Haddix) Hurkley, 57, of Florence Ariz., went to be with her Lord on Wednesday, Feb. 09, 2011, after a short illness. She was born in Midland (Michigan) in 1953 to Otto and Hazel (Dean) Haddix, one of four children. She was predeceased by her parents and one brother, James L. Haddix of Ann Arbor.*
>
> *Lynne married Forrest D. Hurkley on April 17, 1982, in Michigan and lived here for 16 years before moving to Mesa, Ariz., and later to Florence to continue her teaching career. She attended Michigan Technological University majoring in medical technology. She later attended Eastern Michigan University and obtained a bachelor of science degree in biology with a minor in chemistry. She also obtained a master of arts degree from*

Central Michigan University to teach the emotionally impaired. Her years of teaching El students were especially rewarding and seemed to be her special calling.

Lynne was a gifted teacher, much loved and respected by her students and fellow teachers. Over the course of her teaching career, both in Michigan and Arizona, she touched many lives with her outgoing personality and love for her students. She also had a special fondness for animals of all kinds, especially her dogs and cats. She was an accomplished horsewoman in her early years and enjoyed many outdoor activities like hunting, fishing and rock hounding.

Like her mother, she was self-taught in her many hobbies, including quilting and genealogy. She was a woman with "true grit."

Lynne is survived by her husband, Woody Hurkley; one brother, Lee (Susan) Haddix; a sister, Lori Haddix both of Midland, one stepchild, Scott Hurkley of Tucson, Ariz.; and two step grandchildren; along with many nieces and nephews.

The funeral mass will be celebrated today at Assumption of the Blessed Virgin Mary Catholic Church in Florence, Arizona.

Eleven

The 8th Visit to The Farm

I visited The Farm on August 10, 2011. I first took a picture of the house foundation and of the gambrel roof outbuilding. I then went into the barn to the portal location and found a placement of deer hide with fur on the portal. It appeared that someone from the Assinsnaabe had placed an offering for the energies.

Twelve

The Demolition of The Farm

Dan eventually ended up going back to The Farm. Here is what he wrote about what prompted him.

> *I had caught wind that it was being demolished. I was out for a ride, so I thought I would check it out. There was no way I was going on that property again, so I pulled into the neighbor's driveway a few hundred yards away. That is when I got the picture of the black things in the field, even though I could not see them with the naked eye. It did make me sad to see it all in a big pile of broken rubble.*
>
> *I think there are a few reasons the place was destroyed. The first reason might just be to be able to farm that area where the buildings were. Or they might have worried about the liability aspect.*
>
> *However, they were still using the buildings as storage for their equipment. I have experienced spirit activity at and around all the buildings. I wonder if the owner thought that by demolishing the buildings it would rid the area of spirit activity. I don't know if that worked ,but I'm not going back to find out. If they are still there, I hope they can have a peaceful existence.*

On November 14, 2020 I received this email from Dan about the demolition of The Farm.

Hi George,
They were not happy! They made it very difficult to get the shots. I will call you Sunday to fill you in on details.
Dan

Dan sent many pictures after the demolition. Apparently, the energies were displeased with the demolition to the point that Dan had difficulty taking pictures on his iPhone 6 and Canon point and shoot camera. I had received many pictures from Dan of the demolition of The Farm, and I had made mention of the appearance of what were birds in three of the pictures. I then received a call from Dan. We discussed the problems he had taking the pictures and of the 'birds' that appeared in the pictures he took on the south side of The Farm site. He didn't remember any birds when he took the pictures. I received this message from him asking about enhancing the pictures with birds.

Hi George,
Just a thought here. Can you enhance any of the photos I sent you? Not sure if you have software that might clean up or enhance the shots to try to tell what we are looking at (regarding the birds).
Dan

I replied to Dan.

I am doing just that with a message to you, Larry and Caryn. I will devote a paragraph with my thoughts about the three pictures that we thought were birds. They definitely aren't birds, but the graininess of the pictures makes it hard to do a lot with them. My suggestion will be for someone to retake the pictures with better results. Perhaps that person may be able to see what is actually there. My bottom line will be that they appear to be corn shocks or other related field features.

Unfortunately, no one was able to take pictures. I then sent this email to the group.

Hello fellow interested people,
I will be treating this email as a documentary because it will probably be one of the last chapters for the book about The Farm. There are many pictures Dan shared with me that he took on Saturday Nov 14, 2020.
On Friday, the previous day, Dan called me and said that the buildings at The Farm had been demolished and put into piles of rubble. Dan sent me 35 pictures of which I am sharing 18 with you. After he sent the 35 pictures to me, he said "they were not happy." I thought Dan meant the demolition crew was unhappy Well, when Dan called the next day, he said the energies were not happy. (Duh!)
Dan said he charged his iPhone and Canon point and shoot camera before he visited The Farm. As you will see in the pictures, he walked to many vantage points to get his shots that included several that were taken relatively close to the house and barn. Dan had several problems with focus, non-picture activation and battery power levels with both devices. In fact, when he was outside his vehicle and trying to take some pictures, his camera and/or phone showed very low battery levels. When he got back in his vehicle, the device(s) showed that they were at full power. This seems to indicate to me that the metal vehicle he was in shielded his devices from outside energies.

I picked 17 pictures for Caryn, Larry and Gail to let them know the status of The Farm. I comment on each one for reference.

Here are Dan's pictures 1 through 6 with my directional comments about each picture:

1. Looking east from the driveway entrance with the house foundation on the left. 2. Looking southeast from Meridian Rd on or near the bridge over Bullock Creek with the house on the right. 3. Zooming in from picture. 4. Looking northeast from about 100 yards south of the property. 5. Looking east from the north side of the house foundation. Metal scrap pile on the left, barn pile center, other buildings pile on the right. 6. Looking north by northeast at the south side of the house foundation. (Notice the distortion of the side mirror. Are the energies distorting the picture?)

Dan's pictures 10 through 17.

10. The southeast corner of the house foundation as you look northeast. 11. Another view of picture 10 from a looking more from the east. 12. Looking southeast at the southwest corner of the house foundation. 13. The barn pile. 14. The outbuildings pile east of the house foundation. 15. The barn pile looking generally southeast. 16. The barn pile looking northeast. 17. The scrap metal pile.

I sent this message to the group.

> *Well, where do we go now? For me, I will continue on the book. Ideally, I would like to see a return visit with me plus Caryn and Larry to find what the resident energies feel about the change. I am guessing the Native Americans are less than pleased. They apparently weren't the happiest group to start with. I am wondering about the Doutre & Dean residents too, if any of them are in residence. Gail is not wild about the idea of contacting the owners, and Dan has some apparent reservations too.*

But you know me, I want to know more details. I personally don't fear wayward energies, because when they seem to approach, I have told them that I am not afraid of them and to leave me alone. So far, so good. I do want each of you to bend my ear with lengthy thoughts about this latest change at The Farm. We have wonderful records of what used to be and what is still there. Eventually we will all be gone, but our recollections, investigations and other matters regarding this place need to be preserved as a matter of record.
George

Caryn responded.

I scanned your email. You know me with reading and patience, and then I looked at the pics. I looked at them one by one. First very quickly, and two things struck me: one is the negative energy that was weighing down my chest as I scanned them. I looked at them a second time and started noticing 'entities' in most of them.
Definitely they are feeling displaced and not happy at all about this displacement. I then reread the email more thoroughly and looked at the pics one by one as I read George's thoughts. I continued to see the disruptions of the scene, meaning the blurry images. Picture 18 popped out at me the most. Notice the trees, how they seem to want to protect the pile of metal.

Picture 18

The picture of the foundation was the only one I felt peace and joy with. I believe it's because that is where the family was that I helped release to their place of rest.

The house foundation

The pile of lumber holds the 'migrant' labor's souls, I feel that with all certainty. They are perplexed but not angry. Remember how I said they seemed to party there? It was their home.

The Barn Pile

Caryn continued.

The 'birds' when we were there, they served to warn me to stay away from where the Native Americans were, and that is what I feel when I look at them. They too are lost; they have protected the burial grounds for so long that they feel helpless now. They all need to be released. I know the migrant workers could be convinced to go but not sure about the Native Americans. I am going to talk to a few of my Native American friends and see what they think. I have been in a few different Natural American burial grounds with the huts and all seems peaceful there. I am wondering what the story is there for the grounds at The Farm. I think that will be a key to understanding.

These are my off-the-cuff thoughts for now. I intend on sitting in a quiet morning when I get up early by myself and concentrate on each picture with

a notepad next to me. I will send them to you.
Loves to all, Caryn Ann

I sent this message to the group when Dan visited the site for the 10th visit and the piles were removed.

Dan sent me pictures of The Farm location after the piles and other landmarks had been removed. As you can see, there is nothing visible left under the thin blanket of snow. When Dan called me yesterday, he had visited the site, taken the pictures and said that he felt no energies that we are all familiar with.

The Farm site January 2021

Although he did not walk the site, I am guessing there must be some energies still at the portal location. I doubt this surface alteration has significantly reduced what is at the portal, but that is my conjecture. Of course, our documentation of our interactions are much more valuable now. I intend to keep working on the book so that the information and story don't get lost. Of course, I look forward to all of your comments about this end phase of our story.
George

I sent this message to the group regarding the three 'bird' pictures.

> *At first, when Dan and I speculated about them (before I had a good look at them on my big screen monitor), he and I suspected birds. But Dan said, in his phone call to me, that he didn't remember any birds in the shots. As I look at them on my big monitor, they definitely aren't birds. (In our present reality) I have enlarged and enhanced the second picture, but due to limitations of the quality of the original picture, I could only do so much. I have, of course, many questions about these images. They are either corn plant artifacts or some other optical trick. In all three pictures, it appears that the dark objects are in the same relative location.*
>
> *Also, these objects tend to blend in with the surrounding objects as you view them going to the lower left part of the image. They also diminish in size as they blend in with their surroundings.* (Which is a suprise. Normally, objects are bigger when pictures are taken. Larry and I differ on the perspective.)
>
> *As with so many other pictures where energy images are suspected (historically speaking), the quality is just poor enough where it is difficult to deduce strong evidence for what we suspect, namely, individual energy forms.* (I address these the pictures later in this chapter). *These are images where Dan's camera focus was apparently affected. The reason may be either operator error, camera malfunctioning or energy existence seemingly creating visual evidence. The best way to further determine what they actually are would be for Dan and another photographer to take the same pictures again with different cameras to compare results. Or a return visit may show they are just clumps of dirt.*

The first picture of the birds without processing.

The picture with processing. It is enlarged. Unfortunately, there isn't better detail due to the relatively low size of the file.

These are the pictures of the 'birds' in the first 3 images. The last one doesn't show the 'birds'. The first three images were taken with an iPhone 6 and the last one was taken with a Canon Powershot camera. The second picture was the one which I processed and enlarged.

The Farm, The Migration

Larry had this observation about the 'birds' in the pictures.

> *The energies captured in photographs by Dan Braden, as the excavator piled the wood, metal, and stone remnants of the structures that had been the buildings of The Farm, told the final chapter of the story. If you look closely at the dark blobs, seemingly superimposed on the still photographs, you see they resemble a flock of eclectic birds nervously gathering into a flock to begin a migration. They represent a plump pheasant, a flexing hawk or falcon, starlings, doves, robins, sparrows and finches. Clan groups, gathered unto themselves and then into the general flock.*
>
> *One of the photos shows a perspective view of the 'line of birds' where the closest ones are bigger and then taper back to smaller ones farther away. Notice the heads and beaks are all turned to the north, away from what's left of The Farm. There is a famous photo and painting of the Cherokee Nation being driven out of Georgia that has this same perspective.*
>
> *This place we call 'The Farm' is a crossroad of historical energies. It's the 'prime meridian' line that all Michigan counties are oriented to and*

crosses the head section of the stream (Bullock Creek) *(and major Indian footpaths) that flows to the 'Little Forks' on the Tittabawassee River* (Near Midland, Michigan).

The gathering of birds to flocks to migrate and fly in a unison of motion exhibits a quantum energy of motion. The quantum energies of The Farm, the ones that were there when it was built, and the ones that were there before it was built, have flown away.

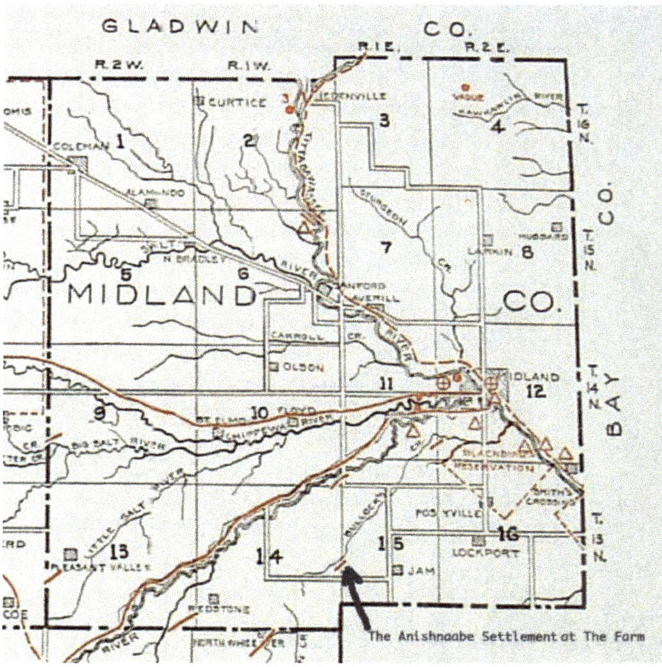

The Anishaabe Settlement at The Farm from a historical map showing Native American trails.

The Farm site October 19, 2021 looking east from the driveway. The house was on the near center left and the barn with the portal was in the area where the field turns tan beyond the dirt.

The barn August 11, 2011.

Thirteen

Evidence Regarding Events

The first event in **Chapter One**, Dan's levitation, is a key event for many reasons. The apparent event of levitation runs counter to classical physics. It does not follow (to me) the principles that Carl Sagan taught about confirming events as true with evidence with our conception of perception of reality. Dan's perception of non-classic physics, when he experienced his event, flies in the face of current reality experiences for most people.

There may a complex understanding of events around how he experienced this event. The physics of this phenomenon do not pass the test of regular critique for most of us. If you're like me, this event doesn't obviously pass the test of reality. But apparently, it happened in Dan's reality.

Was Dan hallucinating? I'm going old school here, reminding us that Dan is a no-nonsense person. He does a have an excellent sense of humor, but my observations of him is that he is a very grounded person. If that is so, he experienced, I suspect, a quantum event that crossed between 'their' parallel dimension and ours.

'Their' level of existence, I suspect, is an energy level related. The energy to lift at least 100 pounds for a distance of about 200 feet at a height of about six inches requires significant amounts of power, and that doesn't make classical physics sense. That event, if it happened, is an enlightened reality check for me. By the circumstantial evidence I've experienced, I am convinced that the levitation happened. However, I need validation through scientific evidence. So, I am about 60% convinced that the event happened, and

I hold 40% doubt it happened due to classic physical laws. It is my gut instinct that the levitation happened.

Chapter Two is the first visit to The Farm. Caryn's sensitivity to energies is displayed. She identifies individuals, not by name, but as people who held residual energies there. Caryn's sensitivities are key events. If she didn't relate her perceptions of this investigation, it would have gone nowhere. This was a part that compelled me to explore more of what was an interesting investigation. At this point of this investigation, I certainly did not expect what was in store facts-wise. I relied on Caryn's ability to relate what was an unfamiliar subject for me. The evidence is Caryn's and my telling of the events that happened.

Of course, it is anecdotal and word of mouth is not evidence, however, I am convinced that Caryn sensed the energies.

In **Chapter Three,** Gail and I visited The Farm. As she and I leaned into one of the outbuildings, she felt an arm around her waist. This is once again anecdotal, but her reaction was instinctive.

Chapter Four is my retelling of a session with a medium who touched on The Farm subject. Again, it is anecdotal, but I have respect for the medium. She has a definite sensitivity with energies in my experience with her.

Chapter Five is a key event because Larry presented newspaper evidence that the energies had names. The event of producing evidence after the first energies is evidence by itself. Larry presented evidence that the energies were people that experienced tragedy in the past. Caryn's energy sensitivity was proven correct.

Chapter Six recounts Hazel's remembrance of the original tragedy. This recollection confirms the evidence Larry provided on the newspaper clipping.

Chapter Seven has the discovery of the Native American portal. Larry's discovery was made possible with observations from Caryn confirming his discovery. The events regarding evidence are, at best, circumstantial. It also provides an obituary of Hazel who was an original member of the Dean family.

Chapter Eight is an anecdotal account about my dreams regarding The Farm.

Chapter Nine reflects my research of the Prime Meridian of Michigan which runs by The Farm.

Chapter Ten includes Lynne's obituary, who provided me with pictures of the family who lived at The Farm.

Chapter Eleven shows one of my photographs of an offering at the portal. This shows another person sensed the energy of the portal.

Chapter Twelve shows a visit where Dan photographed the demolition of The Farm site. The 'birds' in the pictures he took are interesting. It appears to me that, due to his sensitivities to energies, he was able to capture images of energies that were attracted to him. Of course, the camera he used recorded something. He didn't remember the 'birds when he took the pictures. He apparently attracted energies that he didn't recall seeing in real time, although the camera recorded something. A person could try to recreate this event, but Dan probably wouldn't participate, so the instance wouldn't be the same.

The 'birds' pictures were taken with an iPhone 6. The same perspective was taken with a Canon Powershot camera. The 'birds' don't appear in the Powershot camera. Of course, this is a questionable set of circumstances. The 'bird' pictures are a hard thing to prove evidence of. Photographs nearly always lead people to believe that they prove or refute of the truth in the picture. Given that reasoning, these photographs are not evidence.

Chapter Thirteen is this compilation of evidence.

In Conclusion

Quantum physics, I suspect, is the reason we experienced these discoveries. I will try to shed some light on this concept. Here are key points on Quantum Mechanics.

Waves and particles are the foundations of classic physics. Waves are light, magnetism and gravity. Particles are atoms plus subatomic particles. They are at the smallest level of matter as we know them.

They have been established by scholars for much of human history. Quantum Physics are different from the Classic Physics most often taught in mainstream education. Classic Physics is causal, where Quantum Physics is not. Causal is defined as the relationship between cause and effect, everything has a cause in Classic Physics due to standards established by scholars. Quantum Physics are different in that objects are neither waves nor particles. Quantum Theory particles can move between different energy levels.

Quantum Entanglement means that particles can be linked to each other regardless of their distance from each other. Classic Physics states that objects have fixed properties. Quantum objects are wave objects with variable properties depending on how you *observe* them. Quantum Mechanic particles can interact in multiple states; this is called superposition.

Most scholars don't fully understand Quantum Mechanics, but top physicists have calculated mathematical existence of Quantum Mechanics. One of the key points of Quantum Physics is the existence of parallel universes. That concept doesn't make sense in our reality of our Classic Physics. But that concept occurs in Quantum Physics regarding calculations of scholars. Physicists have mathematically calculated the probability of their existence. Eleven separate parallels have been calculated to exist.

This concept is bizarre to comprehend in our environment. However, it explains, to a degree, what we experienced with our visitations. The levitation is still hard to understand, but it may have been caused by Quantum Mechanics. Margaret crossing over to another level to another dimension makes sense with a parallel dimension. Apparently, one of the parallel dimensions is relativity quite close to us. It makes sense to me that energies needing healing or some other conversion can move to higher levels of dimensions or other parameters. What that healing is is not known to me. Perhaps a shaman or similarly enabled person could inform me to that qualification. Apparently, the Native Americans knew this.

The role of Quantum Physics has an apparently important part of us understanding this energy detection. This reality is very strange to most of us, but it explains a lot of what we experienced at The Farm.

About the author

George C. Bailey grew up in Michigan and has followed science all his life, inspired early on by the wonder and clarity of Carl Sagan. He now lives with his wife, Gail, on Michigan's Keweenaw Peninsula, where the northern lights frequently set the sky aglow.

His award-winning photography captures the aurora and the rugged landscapes of the Keweenaw, pairing a patient eye with a deep appreciation for the natural world. When he isn't behind the lens, Bailey enjoys researching family history and helping younger family members record, preserve and share the stories that connect generations.

You can learn more about George at: www.facebook.com/george.c.bailey.98

www.ingramcontent.com/pod-product-compliance
Lightning Source LLC
LaVergne TN
LVHW050916210926
848482LV00012B/465
* 9 7 9 8 9 9 4 7 4 6 6 3 9 *